R Recipes

A Problem-Solution Approach

Larry A. Pace

Apress®

R Recipes: A Problem-Solution Approach

ISBN-13 (pbk): 978-1-4842-0131-2

ISBN-13 (electronic): 978-1-4842-0130-5

Managing Director: Welmoed Spahr
Lead Editor: Steve Anglin
Development Editor: Matthew Moodie
Technical Reviewer: Myron Hlynka
Editorial Board: Steve Anglin, Mark Beckner, Gary Cornell, Louise Corrigan, Jim DeWolf, Jonathan Gennick,
    Robert Hutchinson, Michelle Lowman, James Markham, Matthew Moodie, Jeff Olson, Jeffrey Pepper,
    Douglas Pundick, Ben Renow-Clarke, Gwenan Spearing, Matt Wade, Steve Weiss
Coordinating Editor: Melissa Maldonado
Copy Editor: Kimberly Burton-Weisman
Compositor: SPi Global
Indexer: SPi Global
Artist: SPi Global
Cover Designer: Anna Ishchenko

Distributed to the book trade worldwide by Springer Science+Business Media New York, 233 Spring Street, 6th Floor, New York, NY 10013. Phone 1-800-SPRINGER, fax (201) 348-4505, e-mail orders-ny@springer-sbm.com, or visit www.springeronline.com. Apress Media, LLC is a California LLC and the sole member (owner) is Springer Science + Business Media Finance Inc (SSBM Finance Inc). SSBM Finance Inc is a Delaware corporation.

For information on translations, please e-mail rights@apress.com, or visit www.apress.com.

Apress and friends of ED books may be purchased in bulk for academic, corporate, or promotional use. eBook versions and licenses are also available for most titles. For more information, reference our Special Bulk Sales–eBook Licensing web page at www.apress.com/bulk-sales.

Any source code or other supplementary material referenced by the author in this text is available to readers at www.apress.com. For detailed information about how to locate your book's source code, go to www.apress.com/source-code/.

Contents at a Glance

Contents

About the Author

Larry A. Pace is a statistics author, educator, and consultant. He lives in the upstate area of South Carolina in the town of Anderson. Larry earned his PhD from the University of Georgia, majoring in psychometrics and industrial psychology. He has written more than 100 publications, including books, articles, chapters, and book and test reviews. He has worked in academics for many years, as well as in private industry as a personnel psychologist and organization effectiveness manager. He has programmed in a variety of computer languages and scripting languages, including Fortran, Basic, C++, PHP, Python, and ASP.

Larry has won numerous awards for teaching, research, and service. He is currently a Graduate Research Professor at Keiser University, where he teaches doctoral classes in statistics and research methods. He also teaches statistics, research methods, and tests and measurements part-time at Clemson University. When he is not reading or writing about statistics, he likes to tend a small vegetable and herb garden, play his guitar, and cook on the grill. Larry is married to Shirley Pace, and the Paces have four grown children and two grandsons. The Paces volunteer for Meals on Wheels and are avid recyclers as well as pet rescuers. In a long ago time, when college students carried slide rules instead of cell phones, Larry was a member of a famous local rock band, replete with groupies, but he decided to keep music as a hobby and pursue statistics as a profession.

About the Technical Reviewer

Dr. Myron Hlynka has a bachelor's degree in mathematics from the University of Manitoba, and a PhD in statistics from Pennsylvania State University. He holds P. Stat. status from the Statistical Society of Canada. He is currently a professor in the Department of Mathematics and Statistics at the University of Windsor, in Ontario, Canada, where he has taught for almost 30 years. He has authored over 30 refereed publications and has supervised 30 graduate students. He has over 60 article and book reviews in math reviews at MathSciNet.

Dr. Hlynka's research specialty is queueing theory and stochastic models. He maintains a popular web site in queueing theory at `http://web2.uwindsor.ca/math/hlynka/queue.html`.

Acknowledgments

I love working with Apress. My editorial team consisting of Steve Anglin, Matthew Moodie, Melissa Maldonado, and Kimberly Burton-Weisman kept me on my toes and made my life easier at the same time. It is my privilege to have Myron Hlynka as a technical reviewer once again. His eagle eyes and great suggestions kept me from embarrassing myself in print. The Apress production team is excellent, and they did a great job as usual with production of final copy and graphics. My friend and fellow statistics geek Nat Goodman was an informal but much appreciated sounding board throughout this whole process.

I am also thankful to my dean, Sara Malmstrom, and my department chair, Sue Adragna, for their support of my writing projects. Finally, I thank my wife Shirley for tolerating my obsessions with writing and statistics with good humor and grace. Finally, I want to thank my many students over the years, too many to mention by name, who have taught me how to be a better statistics teacher.

Introduction

R is an open source implementation of the programming language S, created at Bell Laboratories by John Chambers, Rick Becker, and Alan Wilks. In addition to R, S is the basis of the commercially available S-PLUS system. Widely recognized as the chief architect of S, Chambers in 1998 won the prestigious Software System Award from the Association for Computing Machinery, which said Chambers' design of the S system "forever altered how people analyze, visualize, and manipulate data."

Think of R as an integrated system or environment that allows users multiple ways to access its many functions and features. You can use R as an interactive command-line interpreted language, much like a calculator. Type a command, press Enter, and R provides the answer in the R console. R is simultaneously a functional language and an object-oriented language. In addition to thousands of contributed packages, R has programming features, just as all computer programming languages do, allowing conditionals and looping, and giving the user the facility to create custom functions and specify various input and output options.

R is widely used as a statistical computing and software environment, but the R Core Team would rather consider R an environment "within which many classical and modern statistical techniques have been implemented." In addition to its statistical prowess, R provides impressive and flexible graphics capabilities. Many users are attracted to R primarily because of its graphical features. R has basic and advanced plotting functions with many customization features.

Chambers and others at Bell Labs were developing S while I was in college and grad school, and of course I was completely oblivious to that fact, even though my major professor and I were consulting with another AT&T division at the time. I began my own statistical software journey writing programs in Fortran. I might find that a given program did not have a particular analysis I needed, such as a routine for calculating an intraclass correlation, so I would write my own program. BMDP and SAS were available in batch versions for mainframe computers when I was in graduate school—one had to learn Job Control Language (JCL) in order to tell the computer which tapes to load. I typed punch cards and used a card reader to read in JCL and data.

On a much larger and very much more sophisticated scale, this is essentially why the computer scientists at Bell Labs created S (for *statistics*). Fortran was and still is a general-purpose language, but it did not have many statistical capabilities. The design of S began with an informal meeting in 1976 at Bell Labs to discuss the design of a high-level language with an "algorithm," which meant a Fortran-callable subroutine. Like its predecessor S, R can easily and transparently access compiled code from various other languages, including Fortran and C++ among others. R can also be interfaced with a variety of other programs, such as Python and SPSS.

R works in batch mode, but its most popular use is as an interactive data analysis, calculation, and graphics system running in a windowing system. R works on Linux, PC, and Mac systems. Be forewarned that R is not a point-and-click graphical user interface (GUI) program such as SPSS or Minitab. Unlike these programs, R provides terse output, but can be queried for more information should you need it. In this book, you will see screen captures of R running in the Windows operating system.

According to my friend and colleague, computer scientist and bioinformatics expert Dr. Nathan Goodman, statistical analysis essentially boils down to four empirical problems: problems involving description, problems involving differences, problems involving relationships, and problems involving classification. I agree wholeheartedly with Nat. All the problems and solutions presented in this book fall into one or more of those general categories. The problems are manifold, but the solutions are mostly limited to these four situations.

What this Book Covers

This book is for anyone—business professional, programmer, statistician, teacher, or student—who needs to find a way to use R to solve practical problems. Readers who have solved or attempted problems similar to the ones in this book using other tools will readily concur that each tool in one's toolbox works better for some problems than for others. R novices will find best practices for using R's features effectively. Intermediate-to-advanced R users and programmers will find shortcuts and applications that they may not have considered, as well as different ways to do things they might want to do.

The Structure of this Book

The standardized format will make this a useful book for future reference. Unlike most other books, you do not have to start at the beginning and go through this book sequentially. Each chapter is a stand-alone lesson that starts with a typical problem (most of which come from true-life problems that I have faced, or ones that others have described and have given me permission to share). The datasets used with this book to illustrate the solutions should be similar to the datasets readers have worked with, or would like to work with.

Apart from a few contrived examples in the early chapters, most of the datasets and exercises come from real-world problems and data. Following a bit of background, the problem and the data are presented, and then readers learn one efficient way to solve the problem using R. Similar problems will quickly come to mind, and readers will be able to adapt what they learn here to those problems.

Conventions Used in this Book

In this book, code and script segments will be shown this way:

```
> x <- c(1, 3, 5)
> px <- c(0.5, 0.25, 0.25)
> dist <- sample(x, size = 1000, replace = TRUE, prob <- px)
>
```

Code and R functions written inline will also be formatted in the code style.

When you are instructed to perform a command within the R Console or R Editor by using the (limited) point-and-click interface, the instructions will appear as follows: File ➤ Workspace.

Looking Forward

In Chapter 1, you will learn how to get R, how R works, and some of the basic things you can do with R. You will learn how to work with the R interface and the various windows you will find in R. Finally, you will learn how R deals with missing data, vectors, and matrices.

■ ■ ■

Migrating to R: As Easy As 1, 2, 3

There are compelling reasons to use R. An enthusiastic community of users, programmers, and contributors support R and its evolution. R is accurate, produces excellent graphs, has a variety of built-in functions, and is both a functional language and an object-oriented one. R is completely free and is distributed as open-source software. Here is how to get started. It really is as easy as 1, 2, 3.

Getting R Up and Running on Your System

The current version at the time of this writing was R 3.1.0. A recent version needs to be available on your computer in order for you to benefit from the R recipes you will learn in this book. Many users migrate to R from other statistical packages, while other users migrate to R from other programming languages. Both types of users are in for a bit of a shock. R is a programming language, but very much unlike most other ones. R is not exactly a statistics package, but rather an environment that includes many traditional statistical analyses. This is neither a statistics book nor an R programming book, though we will cover elements of both when solving problems within the recipes contained in this book.

Visit the Comprehensive R Archive Network (`http://cran.us.r-project.org/`); see the screen capture in Figure 1-1. Users of PCs and Macs can download precompiled binary files, whereas Linux users may have to do the compiling on their own. However, many Linux systems have R as part of their distributions, so Linux users may already have R preinstalled (I'll show you how to check this later in this section).

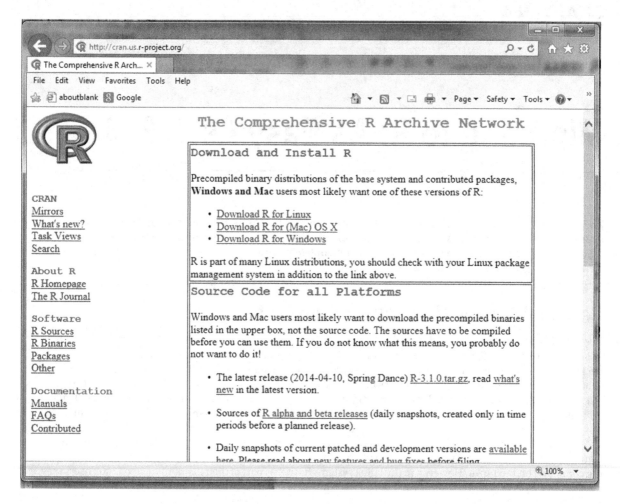

Figure 1-1. *The Comprehensive R Archive Network*

Click **Mirrors** and select the site closest to you. Download the precompiled binary files for your system or follow the instructions for compiling the source code if you need to do so. If you have never installed R, install the base distribution first. Most users of Windows will be able to use the 32-bit version of R. If you want to explore the advantages and disadvantages of using the 64-bit version (assuming you have a 64-bit Windows system), look at the information provided by the R Project to help you choose. You can also do what I did, and install both the 32-bit and the 64-bit versions.

Choose your installation language and options. The defaults are fine for most users. If the R installation was successful, you will have a directory labeled R and a desktop icon for launching R. Figure 1-2 shows the opening screen of R 3.1.0 in a Windows 7 environment.

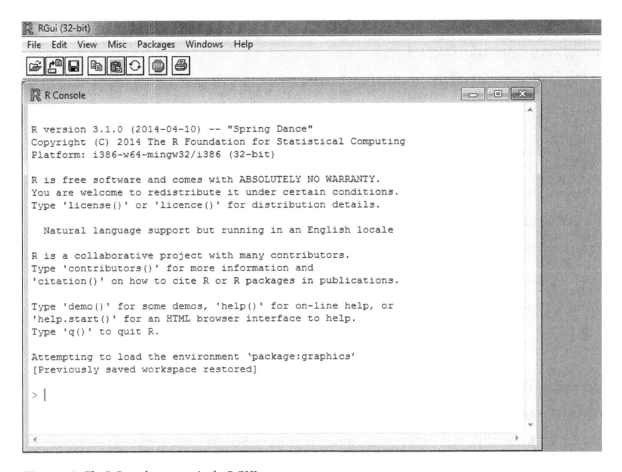

Figure 1-2. *The R Console appears in the R GUI*

As I mentioned, Linux users may have to compile the R source code, but should first check to see if R is distributed with their version of Linux. For instance, I use Lubuntu, a distribution of Linux, on one of my computers, and the base version of R comes prepackaged with Lubuntu, as it does with most Ubuntu versions. To see if you have R base in your Linux system, use the following commands. Open a terminal session. The command prompt in Linux is the tilde character (~) followed by the dollar sign ($).

~$: sudo apt-get install r-base

Once you have installed the base version of R, you can run R from the terminal as follows:

~$: R

Note that the Linux version of R is not likely to be the latest one, as I am currently running R 3.0.2 in Linux (see Figure 1-3) .

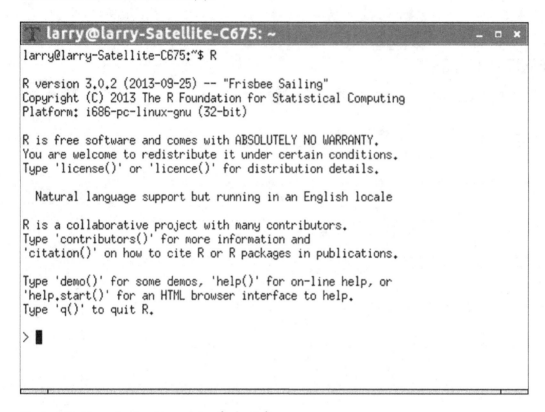

Figure 1-3. *R running in a Linux system (Lubuntu)*

As you see in Figures 1-2 and 1-3, the command prompt in R is >. The following section will show you how to take R for a quick spin.

Okay, So I Have R. What's Next?

Whether you are a programmer or a statistician, or like me, a little of both, R takes some getting used to. Most statistics programs, such as SPSS, separate the data, the syntax (programming language), and the output. R takes a minimalist stance on this. If you are not using something, it is not visible to you. If you need to use something, either you must open it, as in the R Editor for writing and saving R scripts, or R will open it for you, as in the R Graphics Device when you generate a histogram or some other graphic output. So, let's see how to get around in the R interface.

A quick glance shows that the R interface is not particularly fancy, but it is highly functional. Examine the options available to you in the menu bar and the icon bar. R opens with the welcome screen shown in Figure 1-2. You can keep that if you like (I like it), or simply press **Ctrl+L** or select **Edit ➤ Clear Console** to clear the console. You will be working in the R Console most of the time, but you can open a window with a simple text editor for writing scripts and functions. Do this by selecting **File ➤ New script**. The built-in R Editor is convenient for writing longer scripts and functions, but also simply for writing R commands and editing them before you run them. Many R users prefer to use the text editor of their liking. For Windows users, Notepad is fine. When you produce a graphic object, the R Graphics Device will open. The R GUI (graphical user interface) is completely customizable as well.

Although we are showing R running in the R Console, you should be aware that there are several integrated development environments (IDEs) for R. One of the best of these is RStudio.

Do not worry about losing your output when you clear the console. This is simply the view of what you have on the screen at the moment. The output will scroll off the window when you type other commands and generate new output. Your complete R session is saved to a history file, and you can save and reload your R workspaces. The obvious advantage of saving your workspace is that you do not have to reload the data and functions you used in your R session. Everything will be there again when you reload the workspace.

You will most likely not be interested in saving your R workspace with the examples from this chapter. If you do want to save an R workspace, you will receive a prompt when you quit the session. To exit the session, enter **q()** or select **File ➤ Exit**. R will give you the prompt shown in Figure 1-4.

Figure 1-4. *R prompts the user to save the workspace image*

From this point forward, the R Console is shown only in special cases. The R commands and output will always appear in code font, as explained in the introduction. Launch R if it is not already running on your system. The best way to learn from this book is to have R running and to try to duplicate the screens you see in the book. If you can do that, you will learn a great deal about using R for data analysis and statistics.

First, we will do some simple math, and then we will do some more interesting and a little more complicated things. In R, one assigns values to objects with the *assignment operator*. The traditional assignment operator is <-. There is also a little-used right-pointing assignment operator, ->. You can also use the equals sign for assignments. There is some advantage in that you avoid two keystrokes when you use = instead of <-. In this book, we will always use <- for assignments. The = sign is used to specify values for arguments and options in R commands. To test for equality, use ==.

R accepts numbers, characters, variables, and even other functions as input to its functions. R is unlike other languages in several important ways. In most computer languages, a number can be assigned to a constant, usually with an equal sign, =. For example, in Python, you can make the assignment x = 10. The value of 10 is assigned to the variable x. The "type" of x is a scalar quantity (a single value) stored as an integer:

```
Python 3.3.1 (v3.3.1:d9893d13c628, Apr  6 2013, 20:25:12) [MSC v.1600 32 bit (Intel)] on win32
Type "copyright", "credits" or "license()" for more information>>> x = 10
>>> x
10
>>> type(x)
<class 'int'>
```

If you will remember some of your mathematical or computer training, recall that numerical data can be *scalars* (individual values or constants), *arrays* (or vectors) with one row or one column of numbers, or *matrices* with two or more rows and two or more columns. Many computer languages make distinctions among these data types. In some languages, which are called "strongly typed," you must declare the variable's type and dimensionality before you assign a value or values to it. In other languages, known as "loosely typed," You can assign different types of values to the same variable without having to declare the type. R works that way, and is a very loosely typed language.

To R, there are no scalar quantities. When you enter 1 + 1 and then press **Enter**, R displays [1] 2 on the next line and gives you another command prompt. The index [1] indicates that to R, the integer object 2 is an integer vector of length 1. The number 2 is the first (and only) element in that vector. You can assign an R command to a variable (object), say, x, and R will keep that assignment until you change it. When we assign x <- 1 + 1, the value of 2 is assigned to the object x. We can now use x in R commands, such as x + 1. R's indexes start with 1 instead of 0, as some other computer languages do. If you type **numbers <- 1:10**, R will assign the numbers 1 through 10 to the integer vector called *numbers*.

```
> 1 + 1
[1] 2
> x <- 1 + 1
> x + 1
[1] 3
> x * x
[1] 4
> numbers <- 1:10
> numbers
 [1]  1  2  3  4  5  6  7  8  9 10
> numbers ^ 2
 [1]    1    4    9   16   25   36   49   64   81  100
> numbers * x
 [1]  2  4  6  8 10 12 14 16 18 20
> sqrt(numbers)
 [1] 1.000000 1.414214 1.732051 2.000000 2.236068 2.449490 2.645751 2.828427
 [9] 3.000000 3.162278
```

As mentioned at the beginning of this chapter, R is both functional and object-oriented. To R, everything is a function, including the basic mathematics operators. Everything is also an object in R. When you assign x <- 1 + 1, you have created an object called x. One of the most useful and powerful features of R is that many of its operators and functions are *vectorized*

In computer science, something is vectorized if the program works on the vector in elementwise fashion, performing the same operation on each element of the vector that it would have performed on a scalar until it reaches the end of the vector. The general category of array-programming languages includes languages that generalize operations on scalars transparently to vectors, matrices, and higher-order arrays. An operation that works on an entire array is called a *vectorized operation*. Most computer languages are not vectorized to the extent R is. This makes it easy in many situations to avoid explicit loops, which are very slow in comparison to a vectorized operation. If you work in a scientific or engineering setting, you are probably familiar with MATLAB and Octave. Along with R and Python using the NumPy extension, these languages support array programming.

The only other computer language I have worked with that has the same level of vectorization is the now defunct language APL. In most languages, you would have to write a loop to square the numbers from 1 to 10. But in R, you simply use the exponent operator (^) to square all the numbers at once. The primary advantage of this is that you can frequently avoid explicit loops, as mentioned earlier.

R is case sensitive. Note that x and X are different objects in R. Although R is case sensitive, it is insensitive to spaces. I write code that uses spaces and indentation simply to make it easier for me and others to understand, and I usually comment my code fairly liberally. You would be surprised how often you can be doing something that makes perfectly good sense at the time, but looks like total gibberish when you return to it a few months later. Comments help. To insert a comment in a line of R code, simply enter #. The interpreter ignores anything after the # (pound sign or hash tag).

Here's a demonstration of the case sensitivity of R and the use of comments. Instead of working directly in the R GUI, click **File ➤ New Script** to open the R Editor. It is far easier to write and correct multiple lines of code in the editor (or in some other text editor) and execute the code from there than to type directly into the R Console. When you work in the R Editor, leave out the > command prompt. R will supply it (see Figure 1-5).

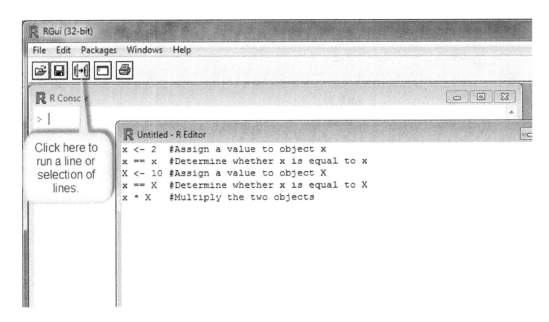

Figure 1-5. *Use the R Editor to write multiple lines of R code*

To execute your code, select one or more lines of code from the R Editor, and then click the icon for running the code in the R Console. As a shortcut, if you want to run all the code, use **Ctrl+A** to select all the code, and then press **Ctrl+R** to run the code in the R Console. Here is what you get:

```
> x <- 2   #Assign a value to object x
> x == x   #Determine whether x is equal to x
[1] TRUE
> X <- 10 #Assign a value to object X
> x == X   #Determine whether x is equal to X
[1] FALSE
> x * X    #Multiply the two objects
[1] 20
>
```

Table 1-1 presents some useful operators, functions, and constants in R.

Table 1-1. *Useful Operator, Functions, and Constants in R*

Operation/Function	R Operator	Code Example
Addition	+	1 + 1
Subtraction	-	2 - 1
Multiplication	*	3 * 2
Division	/	3 / 2
Exponentiation	^	3 ^ 2
Square root	sqrt()	Sqrt(81)
Natural logarithm	log()	> exp(1) [1] 2.718282 > log(exp(1)) [1] 1
Common logarithm	log10()	> log10(100) [1] 2
Complex numbers	complex()	> z <- complex(real = 2, imaginary = 3) > z [1] 2+3i
Pi	pi	> pi [1] 3.141593
Euler's number *e*	exp(1)	> exp(1) [1] 2.718282

Table 1-2 shows R's comparison operators. They evaluate to a logical value of TRUE or FALSE.

Table 1-2. *R Comparison Operators*

Operator	Description	Code Example	Result/Comment
>	Greater than	3 > 2 2 > 3	TRUE FALSE
<	Less than	2 < 3 3 < 2	TRUE FALSE
>=	Greater than or equal to	2 >=2 2 >=3	TRUE FALSE
<=	Less than or equal to	2 <= 2 3 <= 2	TRUE FALSE
==	Equal to	2 == 2 2 == 3	TRUE FALSE
!=	Not equal to	2 != 3 2 !=2	TRUE FALSE

Table 1-3 shows R's logical operators.

Table 1-3. *Logical Operators in R*

Operator	Description	Code Example	Result/Comment

&	Logical And	```> x <- 0:2```
		```> y <- 2:0```
		```> (x < 1) & (y > 1)```
		```[1] TRUE FALSE FALSE```
			This is the vectorized version. It compares two vectors element-wise and returns a vector of TRUE and/or FALSE.

&&	Logical And	```> x <- 0:2```
		```> y <- 2:0```
		```> (x < 1) && (y > 1)```
		```[1] TRUE```
			This is the unvectorized version. It compares only the first value in each vector, left to right, and returns only the first logical result.

|	Logical Or	```> (x < 1)	(y > 1)``` ```[1] TRUE FALSE FALSE```	This is the vectorized version. It compares two vectors element-wise and returns a vector of TRUE and/or FALSE.
||	Logical Or	```> (x < 1) || (y > 1)``` ```[1] TRUE```	This is the unvectorized version. It compares two vectors and returns only the first logical result.
!	Logical Not	```> !y == x``` ```[1] TRUE FALSE TRUE```	Logical negation. Returns either a single logical value or a vector of TRUE and/or FALSE.

Understanding the Data Types in R

As the preceding discussion has shown, R is strange in several ways. Remember R is both functional and object-oriented, so it has a bit of an identity crisis when it comes to dealing with data. Instead of the expected integer, floating point, array, and matrix types for expressing numerical values, R uses vectors for all these types of data. Beginning users of R are quickly lost in a swamp of objects, names, classes, and types. The best thing to do is to take the time to learn the various data types in R, and to learn how they are similar to, and often very different from, the ways you have worked with data using other languages or systems.

R has six "atomic" vector types, including *logical, integer, real, complex, string* (or character) and *raw*. Another data type in R is the list. Vectors must contain only one type of data, but lists can contain any combination of data types. A data frame is a special kind of list and the most common data object for statistical analysis. Like any list, a data frame can contain both numerical and character information. Some character information can be used for factors, and when that is the case, the data type becomes numeric. Working with factors can be a bit tricky because they are "like" vectors to some extent, but are not exactly vectors. My friends who are programmers think factors are "evil," while statisticians like me love the fact that verbal labels can be used as factors in R, because such factors are self-labelling. It makes infinitely more sense to have a column in a data frame labelled sex with two entries, male and female, than it does to have a column labelled sex with 0s and 1s in the data frame.

In addition to vectors, lists, and data frames, R has language objects including *calls, expressions,* and *names*. There are *symbol objects* and *function* objects, as well as *expression* objects. There is also a special object called NULL, which is used to indicate that an object is absent. Missing data in R are indicated by NA.

We next discuss handling missing data. Then we will touch very briefly on vectors and matrices in R.

Handling Missing Data in R

Create a simple vector using the c() function (some people say it means *combine*, while others say it means *concatenate*). I prefer "combine" because there is also a cat() function for concatenating output. For now, just type in the following and observe the results. The na.rm = TRUE option does not remove the missing value, but simply omits it from the calculations.

```
> x <- c(10, NA, 10, 25, 30, 15, 10, 18, 16, 15)
> x
 [1] 10 NA 10 25 30 15 10 18 16 15
> mean(x)
[1] NA
> mean(x, na.rm = TRUE)
[1] 16.55556
>
```

Working with Vectors in R

As you have learned, R treats a single number as a vector of length 1. If you create a vector of two or more objects, the vector must contain only a single data type. If you try to make a vector with multiple data types, R will coerce the vector into a single type. Chapter 3 covers how to deal with various data structure in more detail. For now, the goal is simply to show how R works with vectors.

Because you know how to use the R Editor and the R Console now, we will dispense with those formalities and just show the code and the output together. First, we will make a vector of 10 numbers, and then add a character element to the vector. R coerces the data to a character vector because we added a character object to it. I used the index [11] to add another element to the vector. But the vector now does not contain numbers and you cannot do math on it. Use a negative index, [-11], to remove the character and the R function as.integer() to change the vector back to integers:

```
> x <- 1:10
> x
 [1]  1  2  3  4  5  6  7  8  9 10
> typeof(x)
[1] "integer"
> x[11] <- "happy"
> x
 [1] "1"     "2"     "3"     "4"     "5"     "6"     "7"     "8"     "9"
[10] "10"    "happy"
> typeof(x)
[1] "character"
> x <- x[-11]
> x
 [1] "1"  "2"  "3"  "4"  "5"  "6"  "7"  "8"  "9"  "10"
> x <- as.integer(x)
> x
 [1]  1  2  3  4  5  6  7  8  9 10
> typeof(x)
[1] "integer"
>
```

To make the example a little more interesting, let us work with some real data. The following data (thanks to Nat Goodman for the data) represent the ages in weeks of 20 randomly sampled mice from a much larger dataset.

```
> ages
 [1] 10.5714 13.2857 13.5714 16.0000 10.2857 19.5714 20.0000  7.7143 20.5714
[10] 19.2857 14.0000 14.4286 19.7143 18.0000 13.2857 17.2857  5.2857 16.2857
[19] 14.1429  6.0000
> mean(ages)
[1] 14.46428
> typeof(ages)
[1] "double"
> mode(ages)
[1] "numeric"
> class(ages)
[1] "numeric"
```

R stores numeric values that are not integers in double-precision form. We can access individual elements of a vector with the index or indexes of those elements. Remember that most R functions and operators are vectorized, so that you can calculate the ages of the mice in months by dividing each age by 4. It takes only one line of code (shown in bold), and looping is not necessary.

```
> ages[1]
[1] 10.5714
> ages[20]
[1] 6
> ages[3:9]
[1] 13.5714 16.0000 10.2857 19.5714 20.0000  7.7143 20.5714
> months <- ages/4
> months
 [1] 2.642850 3.321425 3.392850 4.000000 2.571425 4.892850 5.000000 1.928575
 [9] 5.142850 4.821425 3.500000 3.607150 4.928575 4.500000 3.321425 4.321425
[17] 1.321425 4.071425 3.535725 1.500000
```

When you perform operations with vectors of different lengths, R will repeat the values of the shorter vector to match the length of the longer one. This "recycling" is sometimes very helpful as in multiplication by a scalar (vector of length 1), but sometimes produces unexpected results. If the length of the longer vector is a multiple of the shorter vector, this works well. If not, you get strange results like the following:

```
> x <- 1:2
> y <- 1:10
> z <- 1:3
> y/x
 [1] 1 1 3 2 5 3 7 4 9 5
> y/z
 [1]  1.0  1.0  1.0  4.0  2.5  2.0  7.0  4.0  3.0 10.0
Warning message:
In y/z : longer object length is not a multiple of shorter object length
```

Working with Matrices in R

In another peculiarity of R, a matrix is also a vector, but a vector is not a matrix. I know this sounds like doublespeak, but read on for further explanation. A matrix is a vector with *dimensions*. You can make a vector into a one-dimensional matrix if you need to do so. Matrix operations are a snap in R. In this book, we work with two-dimensional matrices only, but higher-order matrices are possible, too.

We can create a matrix from a vector of numbers. Start with a vector of 50 random standard normal deviates (z scores if you like). R fills the matrix columnwise.

```
> zscores <- rnorm(50)
> zscores
 [1] -1.19615960  0.95960082  0.50725210 -0.37411224  1.42044733  1.69437460
 [7]  0.51677914 -0.04810441 -1.28024577 -0.48968148  1.28769546  0.93050145
[13]  0.72614070 -0.19306114 -0.56122938  0.77504861 -0.26756380 -1.11077206
[19] -0.60040090 -0.31920172  1.16802977  1.69736349  0.93134640 -1.15182325
[25]  0.12167256 -1.16038178  1.00415819  0.54469494  1.60231699 -0.11057038
[31]  0.01264523  0.57436245  0.54283138 -0.53045053  0.18115294  1.16062792
[37]  0.63649217  0.59524893 -0.52972220  0.45013366  0.31892391 -0.32371074
[43]  0.89716628 -0.15187155  0.25808226  1.73149549  1.36917698 -0.05803692
[49]  0.44942046  1.07708172
```

```
> zmatrix <- matrix(zscores, nrow = 10, ncol = 5)
> zmatrix
            [,1]       [,2]       [,3]       [,4]       [,5]
 [1,] -1.19615960  1.2876955  1.1680298  0.01264523  0.31892391
 [2,]  0.95960082  0.9305014  1.6973635  0.57436245 -0.32371074
 [3,]  0.50725210  0.7261407  0.9313464  0.54283138  0.89716628
 [4,] -0.37411224 -0.1930611 -1.1518232 -0.53045053 -0.15187155
 [5,]  1.42044733 -0.5612294  0.1216726  0.18115294  0.25808226
 [6,]  1.69437460  0.7750486 -1.1603818  1.16062792  1.73149549
 [7,]  0.51677914 -0.2675638  1.0041582  0.63649217  1.36917698
 [8,] -0.04810441 -1.1107721  0.5446949  0.59524893 -0.05803692
 [9,] -1.28024577 -0.6004009  1.6023170 -0.52972220  0.44942046
[10,] -0.48968148 -0.3192017 -0.1105704  0.45013366  1.07708172
>
```

Imagine the five columns are students' standard scores on four quizzes and a final exam. You can specify names for the rows and columns of the matrix as follows:

```
> rownames(zmatrix)<-c("Jill","Nat","Jane","Tim","Larry","Harry","Barry","Mary","Gary","Eric")
> zmatrix
             [,1]       [,2]       [,3]       [,4]       [,5]
Jill  -1.19615960  1.2876955  1.1680298  0.01264523  0.31892391
Nat    0.95960082  0.9305014  1.6973635  0.57436245 -0.32371074
Jane   0.50725210  0.7261407  0.9313464  0.54283138  0.89716628
Tim   -0.37411224 -0.1930611 -1.1518232 -0.53045053 -0.15187155
Larry  1.42044733 -0.5612294  0.1216726  0.18115294  0.25808226
Harry  1.69437460  0.7750486 -1.1603818  1.16062792  1.73149549
Barry  0.51677914 -0.2675638  1.0041582  0.63649217  1.36917698
Mary  -0.04810441 -1.1107721  0.5446949  0.59524893 -0.05803692
Gary  -1.28024577 -0.6004009  1.6023170 -0.52972220  0.44942046
Eric  -0.48968148 -0.3192017 -0.1105704  0.45013366  1.07708172
```

```
> colnames(zmatrix) <- c("quiz1","quiz2","quiz3","quiz4","final")
> zmatrix
            quiz1       quiz2      quiz3       quiz4       final
Jill  -1.19615960  1.2876955  1.1680298  0.01264523  0.31892391
Nat    0.95960082  0.9305014  1.6973635  0.57436245 -0.32371074
Jane   0.50725210  0.7261407  0.9313464  0.54283138  0.89716628
Tim   -0.37411224 -0.1930611 -1.1518232 -0.53045053 -0.15187155
Larry  1.42044733 -0.5612294  0.1216726  0.18115294  0.25808226
Harry  1.69437460  0.7750486 -1.1603818  1.16062792  1.73149549
Barry  0.51677914 -0.2675638  1.0041582  0.63649217  1.36917698
Mary  -0.04810441 -1.1107721  0.5446949  0.59524893 -0.05803692
Gary  -1.28024577 -0.6004009  1.6023170 -0.52972220  0.44942046
Eric  -0.48968148 -0.3192017 -0.1105704  0.45013366  1.07708172
```

Standardized scores are usually reported to two decimal places. Remove some of the extra decimals to make the next part of the code a little less cluttered. Set the number of decimals by using the round( ) function:

```
zmatrix <round(zmatrix, digits = 2)
zmatrix
      quiz1 quiz2 quiz3 quiz4 final
Jill  -1.20  1.29  1.17  0.01  0.32
Nat    0.96  0.z93 1.70  0.57 -0.32
Jane   0.51  0.73  0.93  0.54  0.90
Tim   -0.37 -0.19 -1.15 -0.53 -0.15
Larry  1.42 -0.56  0.12  0.18  0.26
Harry  1.69  0.78 -1.16  1.16  1.73
Barry  0.52 -0.27  1.00  0.64  1.37
Mary  -0.05 -1.11  0.54  0.60 -0.06
Gary  -1.28 -0.60  1.60 -0.53  0.45
Eric  -0.49 -0.32 -0.11  0.45  1.08
```

If you have occasion to fill a matrix rowwise, set the byrow argument to T or TRUE. You can do this as follows.

```
> y <- matrix(x, nrow = 10, ncol = 10, byrow = TRUE)
> y
      [,1] [,2] [,3] [,4] [,5] [,6] [,7] [,8] [,9] [,10]
 [1,]    1    2    3    4    5    6    7    8    9    10
 [2,]   11   12   13   14   15   16   17   18   19    20
 [3,]   21   22   23   24   25   26   27   28   29    30
 [4,]   31   32   33   34   35   36   37   38   39    40
 [5,]   41   42   43   44   45   46   47   48   49    50
 [6,]   51   52   53   54   55   56   57   58   59    60
 [7,]   61   62   63   64   65   66   67   68   69    70
 [8,]   71   72   73   74   75   76   77   78   79    80
 [9,]   81   82   83   84   85   86   87   88   89    90
[10,]   91   92   93   94   95   96   97   98   99   100
```

R uses two indexes for the elements of a two-dimensional matrix. As with vectors, the indexes must be enclosed in square brackets. A range of values can be specified by use of the colon operator, as in [1:2]. You can also use a comma to indicate a whole row or a whole column of a matrix. Consider the following examples.

```
> y[,1:5]
      [,1] [,2] [,3] [,4] [,5]
 [1,]    1    2    3    4    5
 [2,]   11   12   13   14   15
 [3,]   21   22   23   24   25
 [4,]   31   32   33   34   35
 [5,]   41   42   43   44   45
 [6,]   51   52   53   54   55
 [7,]   61   62   63   64   65
 [8,]   71   72   73   74   75
 [9,]   81   82   83   84   85
[10,]   91   92   93   94   95
> y[1:5,]
      [,1] [,2] [,3] [,4] [,5] [,6] [,7] [,8] [,9] [,10]
 [1,]    1    2    3    4    5    6    7    8    9    10
 [2,]   11   12   13   14   15   16   17   18   19    20
 [3,]   21   22   23   24   25   26   27   28   29    30
 [4,]   31   32   33   34   35   36   37   38   39    40
 [5,]   41   42   43   44   45   46   47   48   49    50
> y[5,5]
[1] 45
> y[10,10]
```

[1] 100R can do many useful things with matrices. For example, calculate the variance-covariance matrix by using the var() function:

```
> varcovar <- var(zmatrix)
> varcovar
            quiz1       quiz2       quiz3       quiz4       final
quiz1   1.0544544 0.11489111 -0.3285267  0.3838900  0.19691333
quiz2   0.1148911 0.63790667  0.1006644  0.1074422  0.07846222
quiz3  -0.3285267 0.10066444  1.0574489 -0.0665400 -0.19844667
quiz4   0.3838900 0.10744222 -0.0665400  0.2859656  0.20044222
final   0.1969133 0.07846222 -0.1984467  0.2004422  0.47039556
```

Invert a matrix by using the solve() function:

```
> inverse <- solve(varcovar)
> inverse
             quiz1       quiz2       quiz3       quiz4       final
quiz1   2.23763294 -0.02683957  0.6305501 -3.3937313  0.7799048
quiz2  -0.02683957  1.72182793 -0.2326712 -0.5743348 -0.1293917
quiz3   0.63055010 -0.23267125  1.2358101 -0.9683403  0.7088308
quiz4  -3.39373126 -0.57433478 -0.9683403 10.3613632 -3.3071827
final   0.77990476 -0.12939168  0.7088308 -3.3071827  3.5292487
```

Do matrix multiplication by using the %*% operator. Just to make things clear, the matrix product of a matrix and its inverse is an identity matrix with 1's on the diagonal and 0's in the off-diagonals. Showing the result with fewer decimals makes this more obvious. For some reason, many of my otherwise very bright students do not "get" scientific notation at all.

```
> identity <- varcovar %*% inverse
> identity
               quiz1          quiz2          quiz3          quiz4          final
quiz1   1.000000e+00   5.038152e-18   3.282422e-17   2.602627e-16   5.529431e-18
quiz2  -8.009544e-18   1.000000e+00  -2.323920e-17   1.080679e-16  -4.710858e-17
quiz3  -7.697835e-17   7.521991e-17   1.000000e+00   9.513874e-17  -9.215718e-17
quiz4   1.076477e-16   1.993407e-17   3.182133e-17   1.000000e+00  -4.325967e-17
final  -4.770490e-18  -6.986328e-18  -1.832302e-17   1.560167e-16   1.000000e+00

> identity <- round(identity, 2)
> identity
      quiz1 quiz2 quiz3 quiz4 final
quiz1     1     0     0     0     0
quiz2     0     1     0     0     0
quiz3     0     0     1     0     0
quiz4     0     0     0     1     0
final     0     0     0     0     1
```

Looking Backward and Forward

In Chapter 1, you learned three important things: how to get R, how to use R, and how to work with missing data and various types of data in R. These are foundational skills. In Chapter 2, you will learn more about input and output in R. Chapter 3 will fill in the gaps concerning various data structures, returning to vectors and matrices, as well as learning how to work with lists and data frames.

CHAPTER 2

Input and Output

R provides many input and output capabilities. This chapter contains recipes on how to read data into R, as well as how to use several handy input and output functions. Although most R users are more concerned with input, there are times when you need to write to a file. You will find recipes for that in this chapter as well.

Oracle boasts that Java is everywhere, and that is certainly true, as Java is in everything from automobiles to cell phones and computers. R is not everywhere, but it is everywhere you need it to be for data analysis and statistics.

Recipe 2-1. Inputting and Outputting Data
Problem

To work with data, you need to get it into your R program. You may want to obtain that data from user input or from a file. Once you have done some processing you may want to output some data.

Solution

Besides typing data into the console, you can use the script editor. The output for your R session appears in the R Console or the R Graphics Device. The basic commands for reading data from a file are read.table() and read.csv().

Note   Here CSV refers to comma-separated values.

You can write to a file using write.table(). In addition to these standard ways to get data into and out of R, there are some other helpful tools as well. You can use data frames, which are a special kind of list. As with any list, you can have multiple data types, and for statistical applications, the data frame is the most common data structure in R. You can get data and scripts from the Internet, and you can write functions that query users for keyboard input.

Before we discuss these I/O (input/output) options, let's see how you can get information regarding files and directories in R. File and directory information can be very helpful. The functions getwd() and setwd() are used to identify the current working directory and to change the working directory. For files in your working directory, simply use the file name. For files in a different directory, you must give the path to the file in addition to the name.

The function file.info() provides details of a particular file. If you need to know whether a particular file is present in a directory, use file.exists(). Using the function objects() or ls() will show all the objects in your workspace. Type **dir()** for a list of all the files in the current directory. Finally, you can see a complete list of file- and directory-related functions by entering the command ?files.

To organize the discussion, I'll cover keyboard and monitor I/O; reading, cleaning, and writing data files; reading and writing text files; and R connections, in that order.

Keyboard and Monitor Access

You can use the scan() function to read in a vector from a file or the keyboard. If you would rather enter the elements of a vector one at a time with a new line for input, just type **x <- scan()** and press the **Enter** key. R gives you the index, and you supply the value. See the following example. When you are finished entering data, just hit the **Enter** key with an empty index.

```
> xvector <- scan()
1: 19
2: 20
3: 31
4: 25
5: 36
6: 43
7: 53
8: 62
9: 40
10: 29
11:
Read 10 items
> xvector
 [1] 19 20 31 25 36 43 53 62 40 29
```

Humans are better and faster at entering data in a column than they are at entering data in a row. You may like this way of entering vectors more than using the c() function.

If your data are in a file in the current working directory, you can enter a vector by using the file name as the argument for scan(). For example, assume you have a vector stored in a file called yvector.txt.

```
> scan("yvector.txt")
Read 10 items
 [1] 22 18 32 39 42 73 37 55 34 34
```

The readline() function works in a similar fashion to get information from the keyboard. For example, you may have a code fragment like the following:

```
> yourName <- readline("Type in Your First and Last Name: ")
Type in Your First and Last Name: Larry Pace
> yourName
[1] "Larry Pace"
```

In the interactive mode, you can print the value of an object to the screen simply by typing the name of the object and pressing **Enter**. You can also use the print() function, but it is not necessary at the top level of the interactive session. However, if you want to write a function that prints to the console, just typing the name of the object will no longer work. In that case, you will have to use the print() function. Examine the following code. I wrote the function in the script editor to make things a little easier to control. I cover writing R functions in more depth in Chapter 11.

```
> cubes
function(x) {
print(x^3)
}
```

```
> x <- 1:20
> cubes(x)
 [1]    1    8   27   64  125  216  343  512  729 1000 1331 1728 2197 2744 3375
[16] 4096 4913 5832 6859 8000
```

Reading and Writing Data Files

R can deal with data files of various types. Tab-delimited and CSV are two of the most common file types. If you load the foreign package, you can read in additional data types, such as SPSS and SAS files.

Reading Data Files

To illustrate, I will get some data in SPSS format from the General Social Survey (GSS) and then open it in R. The GSS dataset is used by researchers in business, economics, marketing, sociology, political science, and psychology. The most recent GSS data are from 2012. You can download the data from www3.norc.org/GSS+Website/Download/ in either SPSS format or Stata format.

Because Stata does a better job than SPSS at coding the missing data in the GSS dataset, I saved the Stata (*.DTA) format into my directory and then opened the dataset in SPSS. This fixed the problem of dealing with missing data, but my data are far from ready for analysis yet. If you do not have SPSS, you can download the open-source program PSPP, which can read and write SPSS files, and can do most of the analyses available in SPSS. The point of this illustration is simply that there are data out there in cyberspace that you can import into R, but you may often have to make a pit stop at SPSS, Stata, PSPP, Excel, or some other program before the data are ready for R. If you have an "orderly" SPSS dataset with variable names that are legal in R, you can open that file directly into R with no difficulty using the foreign() package.

When I read the SPSS data file into R, I see I still have some work to do:

```
> require(foreign)
Loading required package: foreign
> gss2012 <- read.spss("GSS2012.sav")
There were 11 warnings (use warnings() to see them)
> warnings()
Warning messages:
1: In read.spss("GSS2012.sav") :
  GSS2012.sav: Unrecognized record type 7, subtype 18 encountered in system file
2: In `levels<-`(`*tmp*`, value = if (nl == nL) as.character(labels) else paste0(labels,  ... :
  duplicated levels in factors are deprecated
3: In `levels<-`(`*tmp*`, value = if (nl == nL) as.character(labels) else paste0(labels,  ... :
  duplicated levels in factors are deprecated
4: In `levels<-`(`*tmp*`, value = if (nl == nL) as.character(labels) else paste0(labels,  ... :
  duplicated levels in factors are deprecated
5: In `levels<-`(`*tmp*`, value = if (nl == nL) as.character(labels) else paste0(labels,  ... :
  duplicated levels in factors are deprecated
6: In `levels<-`(`*tmp*`, value = if (nl == nL) as.character(labels) else paste0(labels,  ... :
  duplicated levels in factors are deprecated
7: In `levels<-`(`*tmp*`, value = if (nl == nL) as.character(labels) else paste0(labels,  ... :
  duplicated levels in factors are deprecated
8: In `levels<-`(`*tmp*`, value = if (nl == nL) as.character(labels) else paste0(labels,  ... :
  duplicated levels in factors are deprecated
```

```
9: In `levels<-`(`*tmp*`, value = if (nl == nL) as.character(labels) else paste0(labels,  ... :
    duplicated levels in factors are deprecated
10: In `levels<-`(`*tmp*`, value = if (nl == nL) as.character(labels) else paste0(labels,  ... :
    duplicated levels in factors are deprecated
11: In `levels<-`(`*tmp*`, value = if (nl == nL) as.character(labels) else paste0(labels,  ... :
    duplicated levels in factors are deprecated
```

Although this dataset with 4820 records and 1067 variables is large by the standards of the majority of researchers, the data are not "big" in the modern sense. As you can see by the preceding warning messages, the next problem is that the data must be cleaned up a bit before I can do any serious data analysis. Dealing with dirty data is a real-world problem that is not sufficiently addressed in most statistics textbooks, in which professors like me make up examples that are easy to work with, and which almost never have missing data. Recipe 2-2 deals with cleaning up data.

■ **Note**   R nearly choked on the GSS data. We will talk about how to handle very large datasets in Chapter 13.

Writing Data Files

The write.table() function is the analog of the read.table() function. The write.table() function writes a data frame. The function cat() can also be used to write to a file (or to the screen), by successive parts. What this means is that you concatenate the arguments to the cat() function, separating them by commas. You can use any R data type for this purpose. The following code illustrates this:

```
> cats <- c("Tom","Felix","Mittens","Socks","Boots","Fluffy")
> ages <- c(12,10,8,2,5,3)
> pets <- data.frame(cats, ages, stringsASFactors = FALSE)
> pets
      cats ages stringsASFactors
1      Tom   12            FALSE
2    Felix   10            FALSE
3  Mittens    8            FALSE
4    Socks    2            FALSE
5    Boots    5            FALSE
6   Fluffy    3            FALSE
> write.table(pets, "myCats")
> cat("Tom\n", file = "catFile")
> cat("Felix\n", file = "catFile", append = TRUE)
> ## verify the file writes by using the file.exists() function
> file.exists("myCats")
[1] TRUE
> file.exists("catFile")
[1] TRUE
```

Recipe 2-2. Cleaning Up Data

Problem

Real-world data often need cleaning. For example, the GSS codebook uses several different codes for missing data. The easiest way to handle the recoding in this particular case is to clean the dataset in SPSS (see Recipe 2-1 for more on GSS). After the cleaning, the data will be more orderly. In many cases, cleaning data in R is more efficient, and in many others, it might be more efficient to use the search-and-replace functionality of a word processor or a spreadsheet program. As always, choose the most appropriate tool from the toolbox. If the dataset is small, you can make minor edits using the R Data Editor, not to be confused with the script editor.

Solution

When you have serious data recoding and cleaning to do (I call it "data surgery"), I suggest you make use of the plyr package in R. Think of a pair of pliers. The plyr package is a SAC (split-apply-combine) tool, and does a great job for such purposes.

To illustrate some real-world data cleaning issues, let us use a manageable (and I hope interesting to you) set of data, compliments of Dr. Nat Goodman. The data consist of various measurements of mutant and normal mice. The mutated mice were created to carry the genome sequence for Huntington's disease. Several different strains of mice were used because inbred mice are as alike genetically as human twins are. For this example, we will work with only two strains of mice.

The following is the head (the first few records) of the mouse data (which we can view with the head() function). Each mouse has a unique identifier, the strain, the nominal genome sequence, and the actual genome sequence. The sequence CAG repeated seven times represents a normal mouse. CAG sequences of 40 or more are associated with Huntington's disease in mice. The other variables are self-descriptive. The age is the mouse's age in weeks. This dataset represents a small portion of a much larger dataset.

As you have seen previously, when you read in CSV files, you do not have to specify that the first row contains the variable names. The "header" is expected in a CSV file. However, many tab-delimited files do not have a row of column headers. If your tab-delimited file does have a row of variable names as the first row, you must set the header option to T or TRUE, as shown in the following code segment.

```
> mouseWeights <- read.table("Mouse_Weights.txt", header = TRUE)
> head(mouseWeights)
  mouse_id strain cag_nominal cag_actual sex    age body_weight brain_weight
1   hd1769     B6        Q111        113   F 3.7143       11.18        0.380
2   hd1777     B6        Q111        137   F 4.0000       12.50        0.434
3   hd1778     B6          WT          7   F 4.0000       13.30        0.406
4   hd1782     B6        Q111        136   F 4.0000       11.66        0.426
5   hd1806     B6          WT          7   M 4.0000       14.33        0.464
6   hd1808     B6        Q111        113   M 4.0000       13.72        0.414
```

When we examine the data, we see that there are some problems. We find that some mice have an impossible body weight of zero grams. Other mice have an equally impossible brain weight of zero grams.

```
summary(mouseWeights)
    mouse_id       strain      cag_nominal       cag_actual        sex             age
 hd1094 :  1    B6 :376      Q111:172     Min.   :  7.00    F:315    Min.   : 3.714
 hd1095 :  1    CD1:268      Q50 :166     1st Qu.:  7.00    M:329    1st Qu.: 8.000
 hd1104 :  1                 Q92 : 51     Median : 48.00             Median :12.143
 hd1107 :  1                 WT  :255     Mean   : 57.93             Mean   :12.138
 hd1109 :  1                              3rd Qu.:113.00             3rd Qu.:16.286
 hd1110 :  1                              Max.   :154.00             Max.   :20.571
```

21

```
(Other):638
   body_weight        brain_weight
Min.   : 0.00    Min.   :0.0000
1st Qu.:21.12    1st Qu.:0.4580
Median :25.65    Median :0.4900
Mean   :27.57    Mean   :0.4925
3rd Qu.:33.00    3rd Qu.:0.5353
Max.   :59.00    Max.   :0.6660
                 NA's   :12
```

Recode the zeros to missing data as follows. You can attach() the data frame to make it easier to access the individual variables without having to type the data frame name each time you access the variable. After you have attached the data frame, you can refer to variables by their names rather than using the $ format. The following commands will assign a missing value code (NA) to every mouse whose brain weight is equal to zero. Remember that we check for equality in R by using two equals signs (==). Note the square brackets that are used as an index to instruct R to locate all the brain weights of zero and reassign NA to them.

```
> attach(mouseWeights)
> brain_weight[brain_weight==0] <- NA
> body_weight[body_weight==0] <- NA
```

Now, just to illustrate the R Data Editor, say you want to simplify the variable names a little further to make the code more compact. To edit the data frame from inside R, simply enter fix(mouseWeights). The R Data Editor opens in the R GUI (see Figure 2-1).

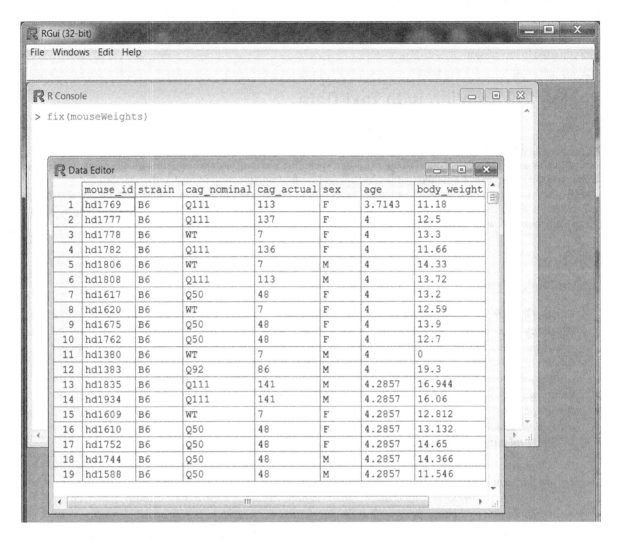

Figure 2-1. The R Data Editor opens in the R GUI

The Data Editor is a simple spreadsheet-like view of your data frame. Make any needed changes, and then when you close the Data Editor, the changes are saved. Here is the newly named set of variables:

```
> head(mouseWeights)
      ID strain CAGnom CAGscale sex     age bodyWt brainWt
1 hd1769     B6   Q111      113   F  3.7143  11.18   0.380
2 hd1777     B6   Q111      137   F  4.0000  12.50   0.434
3 hd1778     B6     WT        7   F  4.0000  13.30   0.406
4 hd1782     B6   Q111      136   F  4.0000  11.66   0.426
5 hd1806     B6     WT        7   M  4.0000  14.33   0.464
6 hd1808     B6   Q111      113   M  4.0000  13.72   0.414
>
```

Finally, tidy things up a bit. Use the detach() function to "unattach" the mouseWeights data. Remove any unneeded objects by using the rm() function, and save the workspace image if you plan to work with the objects and data you used in this session.

Recipe 2-3. Dealing with Text Data
Problem

We are dealing with increasing volumes of text data. Text mining has become an important area of research and innovation, as well as a lucrative one. For our purposes, we define text data as data consisting mostly of characters and words. Text data is typically formatted in lines and paragraphs for human beings to read and understand.

Qualitative researchers treat textual material the same way quantitative researchers treat numbers. Qualitative researchers describe text data, look for relationships and differences, and examine patterns and classifications. There is a growing trend toward combining these methods into a mixed-method research approach.

Solution

Consider Plastic Omnium's environmental policy, which states:

> *Plastic Omnium maintains a proactive environmental protection policy at the highest levels of the company worldwide. It not only ensures compliance with the legal requirements in effect in the countries where Plastic Omnium is present, but in the cases where there are no such requirements or where the company deems the existing requirements inadequate, Plastic Omnium develops and implements its own rules and ensures that they are followed. Every employee involved in an environment-related activity – such as measuring, recordkeeping, composing a report about an action or situation with consequences for the environment, or handling hazardous products or hazardous waste – must take care to perform his or her activities in strict compliance with the laws in effect and only after having received the necessary prior authorizations.*

> *Everyone must ensure that the rules developed by Plastic Omnium are properly applied and will ensure that reports concerning events or situations related to environmental protection are accurate and complete. An employee who is aware of an event or situation within the company, which could result in pollution to the environment, has the duty to take immediate action to bring the matter to the attention of his or her direct supervisor or go directly to the Group's Human Resources Department.*

> —Source: www.plasticomnium.com/en/

Microsoft Word has very rudimentary text analysis tools. We can count the number of words in the policy (there are 205). However, beyond spell checking and grammar checking, there's not too much else we can do using a word processor. R opens up a host of new possibilities.

To do serious text mining in R, you should install the tm package. This topic will be addressed in Chapter 14, but for the present, let's just see how to read the text file into R. I saved the policy as a plain-text file with line feeds only.

```
> Omni <- readLines("Plastic_Omni_Environ_Policy.txt")
> Omni
 [1] "Plastic Omnium maintains a proactive environmental protection policy at the highest levels of "
 [2] "the company worldwide. It not only ensures compliance with the legal requirements in effect in "
 [3] "the countries where Plastic Omnium is present, but in the cases where there are no such "
 [4] "requirements or where the company deems the existing requirements inadequate, Plastic "
 [5] "Omnium develops and implements its own rules and ensures that they are followed. Every "
```

```
 [6] "employee involved in an environment-related activity - such as measuring, recordkeeping, "
 [7] "composing a report about an action or situation with consequences for the environment, or "
 [8] "handling hazardous products or hazardous waste - must take care to perform his or her "
 [9] "activities in strict compliance with the laws in effect and only after having received the "
[10] "necessary prior authorizations."
[11] ""
[12] "Everyone must ensure that the rules developed by Plastic Omnium are properly applied and will "
[13] "ensure that reports concerning events or situations related to environmental protection are "
[14] "accurate and complete. "
[15] ""
[16] "An employee who is aware of an event or situation within the company, which could result in "
[17] "pollution to the environment, has the duty to take immediate action to bring the matter to the "
[18] "attention of his or her direct supervisor or go directly to the Group's Human Resources "
[19] "Department."
[20] ""
```

We use the readLines() function to read in a text file all at once or one line at a time. What is returned is a single character vector. The preceding example reads in a whole file, but if we would rather read in a line at a time, we will have to establish a *connection*. In this case, we will use a connection for file access. Create a connection with various R functions, such as file(), url(), or several additional functions. To see which functions can be used to establish connections, type **?connections** at the command prompt. The parameter r means that we have opened the file for reading. We tell R to read in the lines one at a time by setting the argument n to 1.

```
> connection <- file("Plastic_Omni_Environ_Policy.txt", "r")
> readLines(connection, n = 1)
[1] "Plastic Omnium maintains a proactive environmental protection policy at the highest levels of "
> readLines(connection, n = 1)
[1] "the company worldwide. It not only ensures compliance with the legal requirements in effect in "
>
```

Recipe 2-4. Getting Data from the Internet
Problem

Many datasets are located in repositories on the Internet. There are datasets like the GSS data we have discussed, and literally thousands more web-hosted datasets in economics, data science, finance, government data for the United States and many other countries, health care, machine learning, and various university data repositories. The problem is not so much that we don't have enough data, but instead the problem is that we don't know how to access the right data.

Recipe 2-3 covered how to use a connection to read in a data file line by line. We can also establish a connection to a URL. This makes it possible to read in data from that particular source. The url type of connection supports http://, ftp://, and file://. For additional information on connections, type **?connection** at the R command prompt to see the documentation for the connections() function.

Solution

Recipe 2-3 describes how you can simply copy and paste information from the Internet into a text document and read it into R. However, Recipe 2-4 shows you how to use the scan() function to import a data file. The scan() function, unlike the read.table() function, returns a list or a vector. This makes it easy to read a text file from the Internet. For example, the Institute for Digital Research and Education (IDRE) at UCLA provides excellent R tutorials and example data. Let us read in the file scan.txt from the IDRE web site. We tell R that we want to read the text file into a list with the what argument.

```
> (x <- scan("http://www.ats.ucla.edu/stat/data/scan.txt", what = list(age = 0,
+     name = "")))
Read 4 records
$age
[1] 12 24 35 20

$name
[1] "bobby"    "kate"     "david"    "michael"
```

The read.table() function allows the user to read in any kind of delimited ASCII file. Here's another example from IDRE. In this case, we read in a text file and specify there is a row of column headings by setting the header argument to TRUE.

```
> (test <- read.table("http://www.ats.ucla.edu/stat/data/test.txt", header = TRUE))
   prgtype gender  id ses schtyp level
1  general      0  70   4      1     1
2   vocati      1 121   4      2     1
3  general      0  86   4      3     1
4   vocati      0 141   4      3     1
5 academic      0 172   4      2     1
6 academic      0 113   4      2     1
7  general      0  50   3      2     1
```

CHAPTER 3

Data Structures

As a refresher, the basic data structures in R are vectors, matrices, lists, and data frames. Remember R does not recognize a scalar quantity, instead treating that quantity as a vector of length 1. In Chapter 3, you will learn what you need to know about working with the various data structures in R.

Recipe 3-1. How to Work with Vectors
Problem

Vectors were introduced in Chapter 1, and were described as the fundamental data type in R. In Recipe 3-1, you will learn more about working with vectors, adding and deleting elements, and subsetting vectors. You will also learn more about how vectors relate to other data types in R and how to perform vector operations.

Solution

As you recall, a vector can be any of the six atomic types, but a vector must contain elements of only one data type. As you learned in Chapter 1, you can create a vector from the R Console or the R Editor by entering values with either the c() or the scan() function.

Remember that if you work with vectors of different lengths, R will *recycle* the elements of the shorter vector to match the length of the longer vector. This is often exactly what you want to do, but sometimes it is not. When you use vectors that are *mismatched*, that is, in which the longer vector's length is not a multiple of the shorter vector's length, R will give you a warning to that effect:

```
> x <- 1:10
> y <- 1:5
> x/y
 [1] 1.000000 1.000000 1.000000 1.000000 1.000000 6.000000 3.500000 2.666667
 [9] 2.250000 2.000000
> z <- 1:3
> x/z
 [1]  1.0  1.0  1.0  4.0  2.5  2.0  7.0  4.0  3.0 10.0
Warning message:
In x/z : longer object length is not a multiple of shorter object length
```

Because the length of *x* is a multiple of the length of *y*, the division produced no warning. In the second example, the numbers 1, 2, and 3 were recycled so that on the 10th division, 1 was the element of *z* divided into 10. Next, examine vector arithmetic in R.

As long as the vectors have the same length, all is well. Arithmetic operations work on vectors *elementwise*. That is, the operation is performed for the first element of each vector, then for the second, and so on until the last pair of elements is reached.

```
> xvec
 [1]  0  1  2  3  4  5  6  7  8  9 10 11 12 13
> yvec
 [1]  1  2  3  4  5  6  7  8  9 10 11 12 13 14
> zvec <- xvec + yvec
> zvec
 [1]  1  3  5  7  9 11 13 15 17 19 21 23 25 27
> xvec - yvec
 [1] -1 -1 -1 -1 -1 -1 -1 -1 -1 -1 -1 -1 -1 -1
> xvec * yvec
 [1]   0   2   6  12  20  30  42  56  72  90 110 132 156 182
> xvec / yvec
 [1] 0.0000000 0.5000000 0.6666667 0.7500000 0.8000000 0.8333333 0.8571429
 [8] 0.8750000 0.8888889 0.9000000 0.9090909 0.9166667 0.9230769 0.9285714
```

Vectors are combined by the use of the c() function:

```
> newVec <- c(xvec, yvec)
> newVec
 [1]  0  1  2  3  4  5  6  7  8  9 10 11 12 13  1  2  3  4  5  6  7  8  9 10 11
[26] 12 13 14
```

Remember there is no scalar quantity in R. When you retrieve an element of a vector in R, the result is not really the element itself, but a "vector slice." We use the index or indexes of the vector slice we need by putting the indexes in square brackets ([ ]). Remove an element or elements of a vector by using negative indexes. Add elements or change the values of elements using indexes. If we ask for an element by using an out-of-range index, R will report NA. Let's examine all these operations.

```
> newVec <- newVec[-1]
> newVec
 [1]  1  2  3  4  5  6  7  8  9 10 11 12 13  1  2  3  4  5  6  7  8  9 10 11 12
[26] 13 14
> vecSlice <- newVec[1:13]
> vecSlice
 [1]  1  2  3  4  5  6  7  8  9 10 11 12 13
> vecSlice[14] <- 14
> vecSlice
 [1]  1  2  3  4  5  6  7  8  9 10 11 12 13 14
> vecSlice[15]
[1] NA
```

It is also possible to use a logical index vector to slice a new vector from a given vector. The logical vector must be of the same length as the original vector. The following code illustrates this:

```
> xVec <- 1:10
> logicVec <- c(TRUE,FALSE,TRUE,FALSE,TRUE,FALSE,TRUE,FALSE,TRUE,FALSE)
> vecSlice[logicVec]
[1]  1  3  5  7  9 11 13
```

We can assign names to the elements of vectors. Let us switch to a character vector for this illustration. The names can be used to retrieve and reorder the elements of the vector:

```
> charVec <- c("Phyllis","Argo")
> charVec
[1] "Phyllis" "Argo"
> names(charVec) <- c("FirstName","LastName")
> charVec
FirstName   LastName
"Phyllis"      "Argo"
> charVec[c("LastName","FirstName")]
 LastName FirstName
   "Argo" "Phyllis"
```

The replicate function rep() can be used to create a vector with any number of replications of the same entry:

```
> X <- rep(1,10)
> X
 [1] 1 1 1 1 1 1 1 1 1 1
```

The sequence function seq() requires a *starting value*, an *ending value*, and an *increment value*. For example:

```
> z <- seq(-4, 4, 0.1)
> z
 [1] -4.0 -3.9 -3.8 -3.7 -3.6 -3.5 -3.4 -3.3 -3.2 -3.1 -3.0 -2.9 -2.8 -2.7 -2.6
[16] -2.5 -2.4 -2.3 -2.2 -2.1 -2.0 -1.9 -1.8 -1.7 -1.6 -1.5 -1.4 -1.3 -1.2 -1.1
[31] -1.0 -0.9 -0.8 -0.7 -0.6 -0.5 -0.4 -0.3 -0.2 -0.1  0.0  0.1  0.2  0.3  0.4
[46]  0.5  0.6  0.7  0.8  0.9  1.0  1.1  1.2  1.3  1.4  1.5  1.6  1.7  1.8  1.9
[61]  2.0  2.1  2.2  2.3  2.4  2.5  2.6  2.7  2.8  2.9  3.0  3.1  3.2  3.3  3.4
[76]  3.5  3.6  3.7  3.8  3.9  4.0
```

Recall that missing values are represented in R by NA, and that many R functions will not apply when there are missing data unless you set the argument na.rm to TRUE.

Recipe 3-2. How to Work with Matrices
Problem

As you learned in Chapter 1, R provides many operations for dealing with matrices. We will use real data taken from the General Satisfaction Survey (GSS) for this recipe so that you can see the power of R for working with matrices.

Solution

Matrices and vectors are related, as we have discussed before. A *matrix* is a vector with dimensions. The elements of the matrix must be of the same basic data type. Use the matrix() function to create a matrix. As you will recall, when you create a matrix from data elements, R will fill the matrix *columnwise* by default. Matrix transposition is the interchanging of rows and columns. We accomplish matrix transposition by the function t(). Matrix inversion is done by the solve() function. An illustration of these operations follows.

First, let us extract some variables from the GSS dataset discussed earlier. We will use the cbind() function to create a matrix. Let us use the job satisfaction variable as the dependent variable Y. We will create an Xij matrix by combining a vector of 1s with the age, job security, and income variables. Then we will transpose the Xij matrix and solve for the regression coefficients using matrix operations. You may recall from a statistics class along the way that the column of 1s allow us to calculate the vector of unstandardized regression coefficients. We create our various components as follows. The data frame is a list, as you learned earlier.

```
> head(jobSat)
  age sex race jobsecok income06 satjob7
1  22   1    1        2       25       2
2  36   2    2        2       19       4
3  36   1    1        3       19       3
4  47   2    2        2       18       3
5  54   1    1        3       22       5
6  45   2    3        1       24       2
```

So far, so good. Now for a little matrix wizardry. We create the vector Y, the matrix Xij, and then solve for the regression coefficients using matrix algebra. To explain, the vector Y is simply the column of job satisfaction scores. The matrix Xij is created from a vector of 1s and the age, income, and job security variable. See the following code:

```
> Y <- jobSat$satjob7
> ones <- rep(1, 695)
> Xij <- cbind(ones, jobSat$age, jobSat$race, jobSat$jobsecok, jobSat$income06)
```

With the column of 1s added, our Xij matrix looks like this:

```
> head(Xij)
     index age income06 jobsecok
[1,]     1  22       25        2
[2,]     1  36       19        2
[3,]     1  36       19        3
[4,]     1  47       18        2
[5,]     1  54       22        3
[6,]     1  45       24        1
```

Now, use the traditional matrix formula $B=(X'X)^{-1}X'Y$ to solve for the regression coefficients.

```
> transpose <- t(Xij)
> product <- transpose %*% Xij
> product
          index     age income06 jobsecok
index       695   29451    12569     1094
age       29451 1371371   538411    46474
income06  12569  538411   245013    19675
jobsecok   1094   46474    19675     2118
> inverse <- solve(product)
> inverse
                index           age      income06
index      0.0376712386 -2.949646e-04 -9.507791e-04
age       -0.0002949646  8.236163e-06 -2.714459e-06
income06  -0.0009507791 -2.714459e-06  5.747651e-05
jobsecok  -0.0041537160 -3.148802e-06  1.673929e-05
```

```
                jobsecok
index      -4.153716e-03
age        -3.148802e-06
income06   1.673929e-05
jobsecok   2.531236e-03
> B <- inverse %*% (transpose %*% Y)
> B
              [,1]
index       2.5233
age        -0.0107
income06   -0.0152
jobsecok    0.5213
```

Just for comparison purposes, do this analysis using R's linear model lm() function.

```
> Model <- lm(Y ~ age + income06 + jobsecok)
> summary(Model)

Call:
lm(formula = Y ~ age + income06 + jobsecok)

Residuals:
   Min     1Q Median     3Q    Max
-3.039 -0.811 -0.142  0.611  4.453

Coefficients:
            Estimate Std. Error t value Pr(>|t|)
(Intercept)  2.52326    0.22008   11.47   <2e-16 ***
age         -0.01071    0.00325   -3.29   0.0011 **
income06    -0.01525    0.00860   -1.77   0.0766 .
jobsecok     0.52134    0.05705    9.14   <2e-16 ***
---
Signif. codes:  0 '***' 0.001 '**' 0.01 '*' 0.05 '.' 0.1 ' ' 1

Residual standard error: 1.13 on 691 degrees of freedom
Multiple R-squared:  0.126,     Adjusted R-squared:  0.122
F-statistic: 33.2 on 3 and 691 DF,  p-value: <2e-16
```

As you see, the coefficients are the same as the ones we calculated using matrix algebra.

Recipe 3-3. How to Work with Lists
Problem

Lists are another very important data structure in R. The advantage of a list is that it can combine multiple data types. Recall that indexing is done differently for lists than for vectors and matrices. Lists form the basis for objects such as data frames, and are useful for combining mismatched vectors, as you will soon learn.

Solution

You have learned that there are six atomic vector types in R. A list is a vector, too, but unlike the atomic vectors, which cannot be broken down any further, lists are a special kind of *recursive* vector. Here is a common application for a list. If you are familiar with Python, you will immediately think of a dictionary. In Recipe 3-3, you will learn how to work with lists, including how to create a list, how to access list components and values, and how to apply functions to lists.

Creating a List

Use the list() function to create a list. You might think of a list as a generic vector that can contain other objects. For illustrative purposes, I will create an inventory of a few of the books lying around my desk and in the nearby bookshelves. I include the title, the year of publication, the author, and the publisher. This is the kind of bibliographic information one might be interested in when creating a reference list. First, I entered the information for one of my favorite books.

```
> book <- list(title="Exploratory Data Analysis", year=1977,author="John W. Tukey")
> book
$title
[1] "Exploratory Data Analysis"

$year
[1] 1977

$author
[1] "John W. Tukey"
```

Note that it is not necessary to add component names (also known as *tags*), but they are helpful. We can use the names to retrieve list components. Recall that we index the elements (or components) of a list by using bracket notation, but we can do so in two different ways. We can use either single square brackets ([ ]) or double square brackets ([[ ]]), and the results will be different. Using single brackets results in a list, whereas using double brackets results in a component, and the result will have the type of that component. To illustrate, see that we have three components. Even though book1 and book2 look the same, they are different types of data. Lists can also contain other lists, and they are combined in the same way vectors are.

```
> book <- list(title="Exploratory Data Analysis", year=1977,author="John W. Tukey")
> book
$title
[1] "Exploratory Data Analysis"

$year
[1] 1977

$author
[1] "John W. Tukey"

> book$title
[1] "Exploratory Data Analysis"
> book$year
[1] 1977
> book$author
```

```
[1] "John W. Tukey"
> book1 <- book [1]
> book2 <- book[[1]]
> book1
$title
[1] "Exploratory Data Analysis"

> book2
[1] "Exploratory Data Analysis"
> typeof(book1)
[1] "list"
> typeof(book2)
[1] "character"
> book2 <- list(title="Statistics for the Social Sciences",year=1973,author="William L. Hays")
> books <- c(book1, book2)
> books
$title
[1] "Exploratory Data Analysis"

$title
[1] "Statistics for the Social Sciences"

$year
[1] 1973

$author
[1] "William L. Hays"
```

Adding and Deleting List Components

To add a component to an existing list, simply assign it using a new name and value, or add a list element by using vector indexing:

```
> newList <- list(a = 1, b = 2, c = 3)
> newList
$a
[1] 1

$b
[1] 2

$c
[1] 3

> newList$d <- 4
> newList
$a
[1] 1
```

```
$b
[1] 2

$c
[1] 3

$d
[1] 4

> newList$e <- 5
> newList[6] <- 6
> newList
$a
[1] 1

$b
[1] 2

$c
[1] 3

$d
[1] 4

$e
[1] 5

[[6]]
[1] 6
```

Recall that for vectors, we simply use a negative index, as in [-3], to remove an element. With lists, the way to delete a list element is to assign the special value NULL to the component. Here is an example. Assume I accidentally entered a line feed at the end of entry 9 in my list. I want to delete that entry but leave the others as they are. When I assign NULL to the 9th entry, it is removed from my list, and the length of the list is reduced accordingly.

```
> Grades <- list("A","B","A","B+","C","F","A-","D","B-
+ ","C+")
> Grades
[[1]]
[1] "A"

[[2]]
[1] "B"

[[3]]
[1] "A"

[[4]]
[1] "B+"
```

```
[[5]]
[1] "C"

[[6]]
[1] "F"

[[7]]
[1] "A-"

[[8]]
[1] "D"

[[9]]
[1] "B-\n"

[[10]]
[1] "C+"
> Grades[[9]] <- NULL
> Grades
[[1]]
[1] "A"

[[2]]
[1] "B"

[[3]]
[1] "A"

[[4]]
[1] "B+"

[[5]]
[1] "C"

[[6]]
[1] "F"

[[7]]
[1] "A-"

[[8]]
[1] "D"

[[9]]
[1] "C+"
> length(Grades)
[1] 9
```

Applying Functions to Lists

The lapply() and sapply() functions can be used to apply R functions to lists. In the following code, I compare the quiz scores for two sections of the same statistics class I am currently teaching. The lapply() function applies the mean() function and returns a list, whereas the sapply() function applies the mean function and returns a vector.

```
> quizzes <- list(sect1 = c(10,18,16,16,16,18,14,18,6,20),
+ sect2 = c(18,16,12,16,16,14,18,18,10,14,20,6,16,16,10,14))
> lapply(quizzes,mean)
$sect1
[1] 15.2

$sect2
[1] 14.625

> sapply(quizzes,mean)
  sect1  sect2
15.200 14.625
> list <- lapply(quizzes,mean)
> typeof(list)
[1] "list"
> vector <- sapply(quizzes,mean)
> typeof(vector)
[1] "double"
```

Recipe 3-4. Working with Data Frames

Problem

You have already learned that the data frame is the most frequently used data structure for statistical analysis, and that a data frame is a kind of list. Like matrices, data frames must be rectangular in that every row and column intersection (cell, if you will) contains a value. Data frames, like all lists, can contain any combination of data types, including integer, numeric, character, and logical. Some character variables are used as factors in statistical analyses.

Let us work with some data I collected concerning graduate students' writing assignments in a management class I taught. Data included the course section, the student's sex, the overall course grade, the grades on the Week 2 writing assignment and the Week 6 writing assignment, whether the student used excessive quotation in each assignment, whether the student was documented to have plagiarized the assignment, whether the student voluntarily submitted the assignment to Turnitin.com, the Turnitin similarity indexes for the Week 2 and Week 6 writing assignments, and the percentage of quoted material in each assignment.

Solution

The data described were archival in nature. As a matter of course, I submitted each student's Week 2 and Week 6 (the final week) written assignments to Turnitin.com. The grades were retrieved from the course gradebook. The data represented three sections of the same online course, with 55 students in all. There were some missing data, as one might expect. Students probably committed more plagiarism than the data indicate, because I did not count suspected plagiarism, but only the specific incidents I could document.

Creating a Data Frame and Accessing Data Frame Elements

Data frames can contain any type of data, including other data frames, but in this book, we will limit ourselves to data frames containing numbers and character strings. Applying the length() function returns the number of variables in the dataset, while applying the same function to one of the variables returns the number of records:

```
> plagiarism <- read.csv("plagiarism.csv")
> length(plagiarism)
[1] 17
> length(plagiarism$Course)
[1] 55
```

You can access data frame elements using matrix-like indexing. You can also use variable names to access individual variables (columns, if you will):

```
plagiarism[1,1]
[1] MFE1135A
Levels: MFE1123A MFE1129A MFE1135A
> plagiarism$Course
 [1] MFE1135A MFE1135A MFE1135A MFE1135A MFE1135A MFE1135A MFE1135A MFE1135A
 [9] MFE1135A MFE1135A MFE1135A MFE1129A MFE1129A MFE1129A MFE1129A MFE1129A
[17] MFE1129A MFE1129A MFE1129A MFE1129A MFE1129A MFE1129A MFE1129A MFE1129A
[25] MFE1129A MFE1129A MFE1129A MFE1129A MFE1129A MFE1129A MFE1129A MFE1123A
[33] MFE1123A MFE1123A MFE1123A MFE1123A MFE1123A MFE1123A MFE1123A MFE1123A
[41] MFE1123A MFE1123A MFE1123A MFE1123A MFE1123A MFE1123A MFE1123A MFE1123A
[49] MFE1123A MFE1123A MFE1123A MFE1123A MFE1123A MFE1123A MFE1123A
Levels: MFE1123A MFE1129A MFE1135A

> lowestGrade <- min(plagiarism$CourseGr)
> lowestGrade
[1] 51.01
```

You can filter data and apply functions to data frame elements. For example, let's see who got the lowest overall course grade.

```
> lowestGrade <- min(plagiarism$CourseGr)
> lowestGrade
[1] 51.01
> hist(plagiarism$CourseGr)
```

Is my reputation as a notoriously easy grader still intact (see Figure 3-1)?

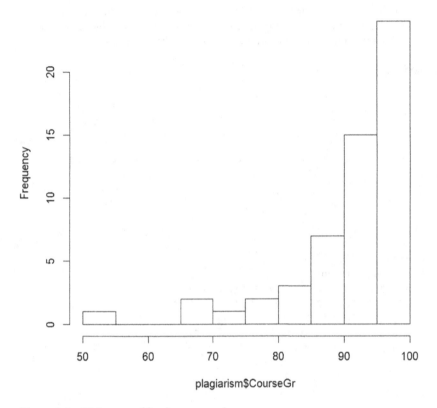

Figure 3-1. *Histogram of final course grades*

Yes, students have nothing to worry about.

Dealing with Missing Data, Take 2

You will recall that in the preparation of the data, missing data must be coded as NA. R will attempt to deal with missing data, but you often must specify the na.rm = TRUE option. If you simply want to remove all records with missing data, use the complete.cases() function.

```
> summary(plagiarism)
     Course    Sex      CourseGr          Wk2Gr               Wk6Gr
MFE1123A:24   F:38   Min.   :51.01   Min.   :  0.00    Min.   :  0.00
MFE1129A:20   M:17   1st Qu.:89.69   1st Qu.: 85.71    1st Qu.: 87.50
MFE1135A:11          Median :93.66   Median : 90.00    Median : 92.50
                     Mean   :90.99   Mean   : 88.00    Mean   : 87.68
                     3rd Qu.:96.72   3rd Qu.: 95.00    3rd Qu.: 97.50
                     Max.   :99.68   Max.   :100.00    Max.   :100.00
```

```
ExcQuote1 Plagiarized1 Plagiarized2 VoluntaryTII         Source1
No :33    No :48       No :54        No :43         Internet    :10
Yes:22    Yes: 7       Yes: 1        Yes:12         Publication  : 2
                                                    StudentPapers: 9
                                                    None         :34


      Wk2Sim          Wk2Inc           PctQuot1          Wk6Sim
Min.   : 0.00   Min.   : 0.00   Min.   : 0.000   Min.   : 0.00
1st Qu.: 4.00   1st Qu.: 8.00   1st Qu.: 0.000   1st Qu.: 3.00
Median : 8.00   Median :17.00   Median : 4.000   Median : 5.00
Mean   :12.73   Mean   :19.98   Mean   : 7.255   Mean   :10.16
3rd Qu.:14.50   3rd Qu.:28.00   3rd Qu.:13.500   3rd Qu.:14.00
Max.   :79.00   Max.   :79.00   Max.   :29.000   Max.   :42.00
                                                 NA's   :10

      Wk6Inc          PctQuot2        ExcQuote2
Min.   : 0.00   Min.   : 0.000   No  :32
1st Qu.: 6.00   1st Qu.: 1.000   Yes :13
Median :10.00   Median : 2.000   NA's:10
Mean   :13.33   Mean   : 3.178
3rd Qu.:17.00   3rd Qu.: 4.000
Max.   :46.00   Max.   :15.000
NA's   :10      NA's   :10
> plagiarism2 <- plagiarism[complete.cases(plagiarism),]
> summary(plagiarism2)
      Course       Sex       CourseGr          Wk2Gr            Wk6Gr
MFE1123A:14   F:31   Min.   :69.99   Min.   :  0.00   Min.   : 70.00
MFE1129A:20   M:14   1st Qu.:89.81   1st Qu.: 85.71   1st Qu.: 87.50
MFE1135A:11          Median :94.52   Median : 90.00   Median : 92.50
                     Mean   :92.16   Mean   : 87.71   Mean   : 91.17
                     3rd Qu.:96.85   3rd Qu.: 94.29   3rd Qu.: 97.50
                     Max.   :99.68   Max.   :100.00   Max.   :100.00
ExcQuote1 Plagiarized1 Plagiarized2 VoluntaryTII         Source1
No :30    No :40       No :44        No :36         Internet    : 7
Yes:15    Yes: 5       Yes: 1        Yes: 9         Publication  : 2
                                                    StudentPapers: 5
                                                    None         :31


      Wk2Sim          Wk2Inc          PctQuot1          Wk6Sim
Min.   : 0.00   Min.   : 2.0   Min.   : 0.000   Min.   : 0.00
1st Qu.: 4.00   1st Qu.: 8.0   1st Qu.: 0.000   1st Qu.: 3.00
Median : 8.00   Median :15.0   Median : 4.000   Median : 5.00
Mean   :12.91   Mean   :19.8   Mean   : 6.889   Mean   :10.16
3rd Qu.:13.00   3rd Qu.:24.0   3rd Qu.: 8.000   3rd Qu.:14.00
Max.   :79.00   Max.   :79.0   Max.   :29.000   Max.   :42.00
      Wk6Inc          PctQuot2        ExcQuote2
Min.   : 0.00   Min.   : 0.000   No :32
1st Qu.: 6.00   1st Qu.: 1.000   Yes:13
Median :10.00   Median : 2.000
Mean   :13.33   Mean   : 3.178
3rd Qu.:17.00   3rd Qu.: 4.000
Max.   :46.00   Max.   :15.000
```

Subsetting Data

You can subset data in many ways. For example, using the dataset *women* from the R distribution, you can select only women who are above the median in weight. The parentheses in the code fragment (women2 <- women[women$weight > 135,]) permits you to save a step and show the selected data immediately.

```
> women
   height weight
1      58    115
2      59    117
3      60    120
4      61    123
5      62    126
6      63    129
7      64    132
8      65    135
9      66    139
10     67    142
11     68    146
12     69    150
13     70    154
14     71    159
15     72    164

summary(women$weight)
   Min. 1st Qu.  Median    Mean 3rd Qu.    Max.
  115.0   124.5   135.0   136.7   148.0   164.0
> (women2 <- women[women$weight > 135, ])
   height weight
9      66    139
10     67    142
11     68    146
12     69    150
13     70    154
14     71    159
15     72    164
```

As you have seen, you can use matrix-type indexing with a data frame. For example, to select only the vector of women's heights, do the following:

```
> (height <- women[,1])
 [1] 58 59 60 61 62 63 64 65 66 67 68 69 70 71 72
```

The subset() function can also be used to select variables using logical tests. To illustrate, let's return briefly to the GSS data first mentioned in Chapter 2, Recipe 2-1. We will select only subjects who are married (marital status = 1).

```
> married <- subset(jobSat, marital == 1)
```

We can also combine logical tests. Select people age 21 and older who worked at least 40 hours last week. Note the use of & in the following.

```
Time40plusHrs <- subset(married, age >=21 & hrs1 >=40)
> head(fullTime40plusHrs)
    wrkstat hrs1 marital age sex race happy weekswrk jobsec jobsecok happy7
24        1   40       1  45   2    3     1       52      2        1      2
35        1   40       1  42   2    1     2       52      5        2      3
75        1   50       1  46   1    1     2       52      5        2      2
83        1   55       1  44   1    1     1       52      5        2      2
118       1   40       1  40   2    1     2       52      4        2      3
127       1   45       1  53   1    1     2       40      2        2      1
    satjob7 satfam7 realrinc  conrinc
24        2       2  49000.0  76600.0
35        3       3  22050.0  34470.0
75        4       2  49000.0  76600.0
83        3       2  33075.0  51705.0
118       2       2  40425.0  63195.0
127       2       1 341672.4 324512.3
```

Saving Datasets

You have already learned that you can write data to various file types, such as CSV and tab-delimited text files. You can also save your datasets in the native R data format, which is the *.rda format. Instead of reading *.rda files, you load them as you would any other R object. This makes it very convenient for other R users who want to work with your data. Let's see how this is done using the data from the plagiarism study. An advantage of loading an R dataset in *.rda format is that it will stay in your workspace when you save it, and you will not have to reload the data during the next session.

```
> plagiarism <- read.csv("plagiarism.csv")
> save(plagiarism, file = "plagiarism.rda")
> file.exists("plagiarism.rda")
[1] TRUE
> load("plagiarism.rda")
> head(plagiarism)
   Course Sex CourseGr Wk2Gr Wk6Gr ExcQuote1 Plagiarized1 Plagiarized2
1 MFE1135A   F    84.55 92.86  97.5        No           No           No
2 MFE1135A   F    91.29 87.14  90.0        No           No           No
3 MFE1135A   M    89.83 84.29  80.0       Yes           No           No
4 MFE1135A   M    91.43 87.14  77.5       Yes          Yes           No
5 MFE1135A   F    71.20  0.00  70.0       Yes          Yes          Yes
6 MFE1135A   F    96.60 88.57  97.5       Yes          Yes           No
  VoluntaryTII        Source1 Wk2Sim Wk2Inc PctQuot1 Wk6Sim Wk6Inc PctQuot2
1           No           <NA>     10     10        0     11     14        3
2           No           <NA>      4     18       14     21     32       11
3           No       Internet      5     31       26      4      9        5
4           No       Internet     11     11        0     18     19        1
5           No    Publication     62     63        1     13     28       15
6           No StudentPapers     55     55        0      2      6        4
  ExcQuote2
1        No
2       Yes
3       Yes
4        No
5       Yes
6        No
```

■ ■ ■

## CHAPTER 4

Merging and Reshaping Datasets

This chapter covers how to merge datasets, add rows and columns to existing datasets, reshape datasets, and stack and unstack datasets. You will find that some of the analyses you want to do will require stacked data, others will require unstacked data, and still others can use data of either type.

Recipe 4-1. Merging Datasets by a Common Variable
Problem

We often have datasets with one or more common variables and want to combine those datasets by matching on a common variable. The merge() function locates matching variables and combines datasets based on these variables.

Solution

The following hypothetical data represent the information on 20 students and each student's scores on five quizzes along with the student's final grade (the average of the quiz scores). See that the only variable the data frames have in common is the student number in column 1.

```
> studentInfo
   Student    Sex Age
1        1   male  18
2        2   male  19
3        3   male  17
4        4   male  20
5        5 female  23
6        6 female  18
7        7   male  21
8        8 female  20
9        9 female  23
10      10 female  21
11      11 female  23
12      12   male  18
13      13   male  21
14      14   male  17
15      15   male  19
16      16 female  20
17      17 female  19
```

```
18        18 female  22
19        19 female  22
20        20   male  20
> studentQuizzes
   Student     Sex Age Quiz1 Quiz2 Quiz3 Quiz4 Quiz5 FinalGrade
1        1    male  18    83    87    81    80    69       69.7
2        2    male  19    76    89    61    85    75       67.5
3        3    male  17    85    86    65    64    81       66.3
4        4    male  20    92    73    76    88    64       68.8
5        5  female  23    82    75    96    87    78       73.5
6        6  female  18    88    73    76    91    81       71.2
7        7    male  21    89    71    61    70    75       64.5
8        8  female  20    89    70    87    76    88       71.7
9        9  female  23    92    85    95    89    62       74.3
10      10  female  21    86    83    77    64    63       65.7
11      11  female  23    90    71    91    86    87       74.7
12      12    male  18    84    71    67    62    70       62.0
13      13    male  21    83    80    89    60    60       65.5
14      14    male  17    79    77    82    63    74       65.3
15      15    male  19    89    80    64    94    78       70.7
16      16  female  20    76    85    65    92    82       70.0
17      17  female  19    92    76    76    74    91       71.3
18      18  female  22    75    90    78    70    76       68.5
19      19  female  22    87    87    63    73    64       66.0
20      20    male  20    75    74    63    91    87       68.3
```

To merge the datasets, do the following:

```
> studentComplete <- merge(studentInfo, studentQuizzes)

> studentComplete
   Student     Sex Age Quiz1 Quiz2 Quiz3 Quiz4 Quiz5 FinalGrade
1        1    male  18    83    87    81    80    69       69.7
2       10  female  21    86    83    77    64    63       65.7
3       11  female  23    90    71    91    86    87       74.7
4       12    male  18    84    71    67    62    70       62.0
5       13    male  21    83    80    89    60    60       65.5
6       14    male  17    79    77    82    63    74       65.3
7       15    male  19    89    80    64    94    78       70.7
8       16  female  20    76    85    65    92    82       70.0
9       17  female  19    92    76    76    74    91       71.3
10      18  female  22    75    90    78    70    76       68.5
11      19  female  22    87    87    63    73    64       66.0
12       2    male  19    76    89    61    85    75       67.5
13      20    male  20    75    74    63    91    87       68.3
14       3    male  17    85    86    65    64    81       66.3
15       4    male  20    92    73    76    88    64       68.8
16       5  female  23    82    75    96    87    78       73.5
17       6  female  18    88    73    76    91    81       71.2
18       7    male  21    89    71    61    70    75       64.5
19       8  female  20    89    70    87    76    88       71.7
20       9  female  23    92    85    95    89    62       74.3
```

Notice that the student numbers are no longer in the original order. R merges on the student number and then sorts on the common variable. But in this case, the number 1 is followed by 10-19, then 2, 20, and 3-9. We can get the numbers back into order by using the order() function, as follows. Note that the row numbers are still the ones associated with the original records, but the new data frame shows the student numbers in order once again.

```
> studentComplete <- studentComplete[order(studentComplete[1]),]
> head(studentComplete)
   Student    Sex Age Quiz1 Quiz2 Quiz3 Quiz4 Quiz5 FinalGrade
1        1   male  18    83    87    81    80    69       69.7
12       2   male  19    76    89    61    85    75       67.5
14       3   male  17    85    86    65    64    81       66.3
15       4   male  20    92    73    76    88    64       68.8
16       5 female  23    82    75    96    87    78       73.5
17       6 female  18    88    73    76    91    81       71.2
```

It is possible that the variables on which you would like to merge datasets have different names in the different datasets (remember R is case sensitive). To deal with that situation, you can either rename the variables so that the names match, or you can use the by.x and by.y options in the merge() command. As an example, what if the student numbers had different variable names, such as id in one file and studentID in another, as follows:

```
> head(studentInfo)
  studentID    Sex Age
1         1   male  18
2         2   male  19
3         3   male  17
4         4   male  20
5         5 female  23
6         6 female  18
> head(studentQuizzes)
  id    Sex Age Quiz1 Quiz2 Quiz3 Quiz4 Quiz5 FinalGrade
1  1   male  18    83    87    81    80    69       69.7
2  2   male  19    76    89    61    85    75       67.5
3  3   male  17    85    86    65    64    81       66.3
4  4   male  20    92    73    76    88    64       68.8
5  5 female  23    82    75    96    87    78       73.5
6  6 female  18    88    73    76    91    81       71.2
```

The label for the merged column is from the first data frame. See that the sex and age variables were inherited from both data frames and are now labeled by their sources.

```
> newData <- merge(x=studentInfo, y=studentQuizzes, by.x="studentID", by.y ="id")
> head(newData)
  studentID  Sex.x Age.x  Sex.y Age.y Quiz1 Quiz2 Quiz3 Quiz4 Quiz5 FinalGrade
1         1   male    18   male    18    83    87    81    80    69       69.7
2         2   male    19   male    19    76    89    61    85    75       67.5
3         3   male    17   male    17    85    86    65    64    81       66.3
4         4   male    20   male    20    92    73    76    88    64       68.8
5         5 female    23 female    23    82    75    96    87    78       73.5
6         6 female    18 female    18    88    73    76    91    81       71.2
```

To eliminate the duplicated columns, you can set them to NULL, as we have discussed previously, or you can tell R to merge on multiple columns to avoid this problem in the first place. I edited both data frames to use the same names for id, sex, and age. Now I simply merge the two data frames as follows:

```
> merge(studentInfo, studentQuizzes, c("id", "sex", "age"))
   id    sex age Quiz1 Quiz2 Quiz3 Quiz4 Quiz5 FinalGrade
1   1   male  18    83    87    81    80    69       69.7
2  10 female  21    86    83    77    64    63       65.7
3  11 female  23    90    71    91    86    87       74.7
4  12   male  18    84    71    67    62    70       62.0
5  13   male  21    83    80    89    60    60       65.5
6  14   male  17    79    77    82    63    74       65.3
7  15   male  19    89    80    64    94    78       70.7
8  16 female  20    76    85    65    92    82       70.0
9  17 female  19    92    76    76    74    91       71.3
10 18 female  22    75    90    78    70    76       68.5
11 19 female  22    87    87    63    73    64       66.0
12  2   male  19    76    89    61    85    75       67.5
13 20   male  20    75    74    63    91    87       68.3
14  3   male  17    85    86    65    64    81       66.3
15  4   male  20    92    73    76    88    64       68.8
16  5 female  23    82    75    96    87    78       73.5
17  6 female  18    88    73    76    91    81       71.2
18  7   male  21    89    71    61    70    75       64.5
19  8 female  20    89    70    87    76    88       71.7
20  9 female  23    92    85    95    89    62       74.3
```

When you merge data frames, R will exclude any observations that appear in only one dataset. Here are some data from the CIA World Factbook web site. We have the economic and demographic data in a CSV file called demographic.csv, and the number of airports in each country in a separate CSV file called airports.csv. After reading those files in with the read.csv() function, I see that they contain data for different countries. The merge will exclude countries not in both datasets. The demographic information covers 46 countries, 45 of which are also in the airport dataset. The airport information includes 236 countries. The merge includes only the countries in both datasets. The final list of countries included in the merged dataset is shown at the end of the following R code:

```
> head(demographic)
      country      area g20 petroleum population pct65plus lifeExpectancy
1     Algeria 2,381,740   0         2     31,736      4.07          69.95
2   Argentina 2,766,890   1         1     37,385     10.42          75.26
3   Australia 7,686,850   1         1     19,357     12.50          79.87
4     Austria    83,858   0         0      8,150     15.38          77.84
5     Belgium    30,510   0         0     10,259     16.95          77.96
6      Brazil 8,511,965   1         1    174,469      5.45          63.24
  Literacy  GDP labor unempl exports imports cellPhones
1     61.6  5.5   9.1     30    19.6     9.2      0.034
2     96.2 12.9    15     15    26.5    25.2      3.000
3    100.0 23.2   9.5    6.4    69.0    77.0      6.400
4     98.0 25.0   3.7    5.4    63.2    65.6      4.500
5     98.0 25.3  4.34    8.4   181.4   166.0      1.000
6     83.3  6.5    79    7.1    55.1    55.8      4.400
```

```
> head(airports)
        country airports
1 United States   13,513
2         Brazil    4,093
3         Mexico    1,714
4         Canada    1,467
5         Russia    1,218
6      Argentina    1,138
> completeCIA <- merge(demographic, airports)
> head(completeCIA)
    country       area g20 petroleum population pct65plus lifeExpectancy
1   Algeria  2,381,740   0         2     31,736      4.07          69.95
2 Argentina  2,766,890   1         1     37,385     10.42          75.26
3 Australia  7,686,850   1         1     19,357     12.50          79.87
4   Austria     83,858   0         0      8,150     15.38          77.84
5   Belgium     30,510   0         0     10,259     16.95          77.96
6    Brazil  8,511,965   1         1    174,469      5.45          63.24
  Literacy  GDP labor unempl exports imports cellPhones airports
1     61.6  5.5   9.1     30    19.6     9.2      0.034      157
2     96.2 12.9    15     15    26.5    25.2      3.000    1,138
3    100.0 23.2   9.5    6.4    69.0    77.0      6.400      480
4     98.0 25.0   3.7    5.4    63.2    65.6      4.500       52
5     98.0 25.3  4.34    8.4   181.4   166.0      1.000       41
6     83.3  6.5    79    7.1    55.1    55.8      4.400    4,093
> length(completeCIA$airports)
[1] 45
> length(demographic$country)
[1] 46
> length(airports$country)
[1] 236
> completeCIA$country
 [1] Algeria              Argentina            Australia
 [4] Austria              Belgium              Brazil
 [7] Canada               China                Czech Republic
[10] Denmark              Finland              France
[13] Germany              Greece               Hungary
[16] Iceland              India                Indonesia
[19] Iran                 Iraq                 Ireland
[22] Italy                Japan                Kuwait
[25] Libya                Luxembourg           Mexico
[28] Netherlands          New Zealand          Nigeria
[31] Norway               Poland               Portugal
[34] Qatar                Russia               Saudi Arabia
[37] South Africa         Spain                Sweden
[40] Switzerland          Turkey               United Arab Emirates
[43] United Kingdom       United States        Venezuela
46 Levels: Algeria Argentina Australia Austria Belgium Brazil Canada ... Venezuela
```

Recipe 4-2. Adding Rows and Columns

Problem

A common problem is the need to add new rows or columns to a data frame. In Recipe 4-2, you will learn how to do that.

Solution

To add rows and columns of data, you use the rbind() and cbind() functions, respectively. I have a vector containing the weights of 40 adult males who exercise regularly. I want to create a vector of id numbers and combine that vector with the weights. This is a job for the cbind() function. The vector of weights is transformed to a data frame (using the as.data.frame() function), and then the vector of id numbers is bound to the weights, as shown in the following code:

```
> weights
 [1] 169.1 144.2 179.3 175.8 152.6 166.8 135.0 201.5 175.2 139.0 156.3 186.6
[13] 191.1 151.3 209.4 237.1 176.7 220.6 166.1 137.4 164.2 162.4 151.8 144.1
[25] 204.6 193.8 172.9 161.9 174.8 169.8 213.3 198.0 173.3 214.5 137.1 119.5
[37] 189.1 164.7 170.1 151.0
> weights <- as.data.frame(weights)
> id <- c(1:40)
> id
 [1]  1  2  3  4  5  6  7  8  9 10 11 12 13 14 15 16 17 18 19 20 21 22 23 24 25
[26] 26 27 28 29 30 31 32 33 34 35 36 37 38 39 40
> weights <- cbind(weights, id)
)
> > head(weights)
  weights id
1   169.1  1
2   144.2  2
3   179.3  3
4   175.8  4
5   152.6  5
6   166.8  6
```

Now, imagine I obtain the weights of another dozen men who exercise regularly. I used the rnorm() function to create the additional data. The rnorm() function produces a random sample of normally distributed scores. I created a 12-element vector with a mean of 175 pounds (lb.) and a standard deviation of 20. The combination of the two vectors produced a matrix, but it is easy to bind it to the data frame.

```
> extraWeights <- rnorm(12,175,20)
> extraWeights
 [1] 166.8790 191.1003 187.1548 158.1391 136.4888 162.1276 211.9919 189.0800
 [9] 202.9519 205.6483 197.1733 173.3894
> extraWeights <- round(extraWeights, 2)
> extraWeights
 [1] 166.88 191.10 187.15 158.14 136.49 162.13 211.99 189.08 202.95 205.65
[11] 197.17 173.39
> newids <- c(41:52)
> newData <- cbind(extraWeights, newids)
> newData
```

```
     extraWeights newids
 [1,]       166.88     41
 [2,]       191.10     42
 [3,]       187.15     43
 [4,]       158.14     44
 [5,]       136.49     45
 [6,]       162.13     46
 [7,]       211.99     47
 [8,]       189.08     48
 [9,]       202.95     49
[10,]       205.65     50
[11,]       197.17     51
[12,]       173.39     52
```

Next, I made sure the two datasets had the same variable names for the rbind() function to work properly. I used the Data Editor and changed the variable names in newData, as shown in Figure 4-1.

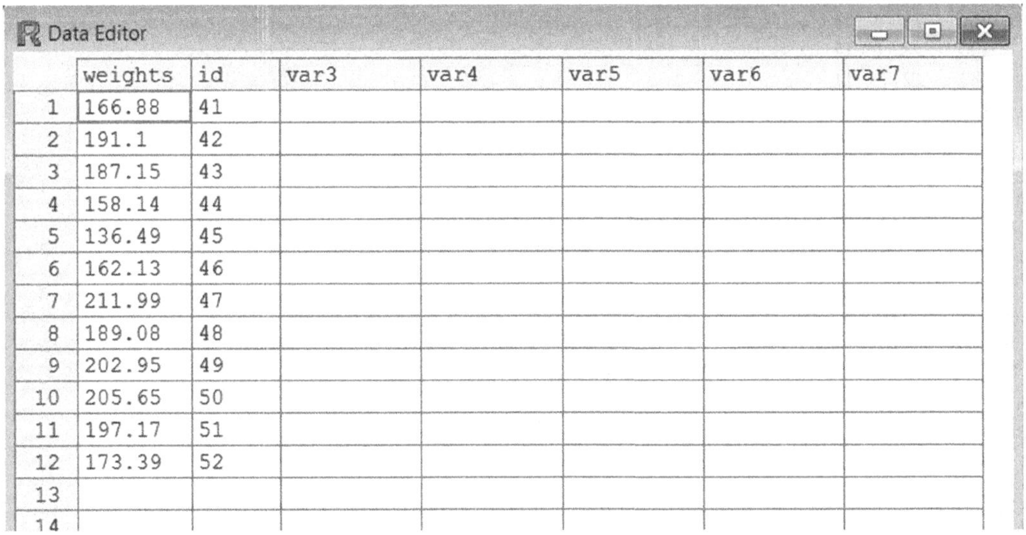

Figure 4-1. *Using the Data Editor to change variable names*

Then, I combined the data using rbind() and verified that all 52 observations were in the new dataset.

```
> weights <- rbind(weights, newData)
> summary(weights)
> length(weights$id)
[1] 52
```

Recipe 4-3. Reshaping a Dataset

Problem

Many repeated-measures datasets have a column for each measurement. For example, here are some repeated-measures data representing the scores on a 20-item test of algebra at the beginning of a statistics course, at the end of the course, and six months later. The *wide* version has each score in a separate column for each student. The *long* version has all 30 scores in a single column, and the time of the test is coded 1, 2, 3 for each measurement. The student numbers are repeated, which means that each student's data occupy 10 rows in the dataset. It's quite common to want to convert from the wide version to the long version, and vice versa.

```
> wideData
   Student Before After SixMo
1        1     13    15    17
2        2      8     8     7
3        3     12    15    14
4        4     12    17    16
5        5     19    20    20
6        6     10    15    14
7        7     10    13    15
8        8      8    12    11
9        9     14    15    13
10      10     11    16     9
> head(longData)
   Case Student Time Score
1     1       1    1    13
2     2       2    1     8
3     3       3    1    12
4     4       4    1    12
5     5       5    1    19
6     6       6    1    10
> tail(longData)
   Case Student Time Score
25   25       5    3    20
26   26       6    3    14
27   27       7    3    15
28   28       8    3    11
29   29       9    3    13
30   30      10    3     9
```

Solution

The reshape() function can convert wide format to long format, and long format to wide format. Let us take the wide data first and convert that dataset to long form. I use separate lines to make the function easier to comprehend:

```
> longScores <- reshape(wideData,
+ direction = "long",
+ varying = list(c("Before","After","SixMo")),
+ times=c(1,2,3),
+ timevar = "Time", idvar = "Student",
+ v.names = "Score")
```

The reshape() function requires several arguments:

- direction tells R what the desired shape of the new data frame will be.

- varying corresponds to the variable names in the wide format that will become separate variables in the reshaped data.

- times indicates the values to use for the newly created time variable specified by the timevar argument.

- v.names gives the names of the variables in the long format that correspond to multiple variables in the wide format.

We have reshaped the wide dataset into a long dataset (see Figure 4-2).

R RGui (32-bit) - [Data Editor]

R File Windows Edit Help

	row.names	Student	Time	Score	var5
1	1.1	1	1	13	
2	2.1	2	1	8	
3	3.1	3	1	12	
4	4.1	4	1	12	
5	5.1	5	1	19	
6	6.1	6	1	10	
7	7.1	7	1	10	
8	8.1	8	1	8	
9	9.1	9	1	14	
10	10.1	10	1	11	
11	1.2	1	2	15	
12	2.2	2	2	8	
13	3.2	3	2	15	

Figure 4-2. *The reshaped data in long form*

The reshape() function does not require the v.names and timevar arguments. They can be used to give appropriate names to the variables.

Now, let's reverse the process. We will convert the long dataset back into a wide dataset. The first order of business is to sort the longScores dataset by student number so that each student's data will occupy three consecutive rows. Changing from one format to the other is often necessary because the data must be in a particular form in order for the analysis to work properly.

```
> head(longScores)
    Student Time Score
1.1      1    1    13
2.1      2    1     8
3.1      3    1    12
4.1      4    1    12
5.1      5    1    19
6.1      6    1    10
```

```
> longScores.sort <- longScores[order(longScores$Student),]
> head(longScores.sort)
    Student Time Score
1.1      1    1    13
1.2      1    2    15
1.3      1    3    17
2.1      2    1     8
2.2      2    2     8
2.3      2    3     7
```

Now, with the data in the correct format, we can change the long data back into wide data, as follows. Note that the time variable and the id variable are the same as in the previous example:

```
> wide <- reshape(longScores.sort,timevar = "Time", idvar = "Student", direction = "wide")
> wide
     Student Score.1 Score.2 Score.3
1.1       1      13      15      17
2.1       2       8       8       7
3.1       3      12      15      14
4.1       4      12      17      16
5.1       5      19      20      20
6.1       6      10      15      14
7.1       7      10      13      15
8.1       8       8      12      11
9.1       9      14      15      13
10.1     10      11      16       9
```

Recipe 4-4. Stacking and Unstacking Data
Problem

Some R functions allow you to use either stacked or unstacked data. For example, here are the scores on a recent 200-point quiz for two sections of my online psychological statistics class. The data are in both stacked and unstacked form. For example, the t.test() function can be used with either stacked or unstacked data, but that is not true of many other analyses, so learning to stack and unstack data as required is a very useful basic R skill.

```
> unstacked
  section1 section2
1      176      120
2      176      199
3       98      159
4      118      127
```

```
5       103      141
6       190      132
7       173      176
8       184       52
9       149      180
> stacked
    score  section
1     176 section1
2     176 section1
3      98 section1
4     118 section1
5     103 section1
6     190 section1
7     173 section1
8     184 section1
9     149 section1
10    120 section2
11    199 section2
12    159 section2
13    127 section2
14    141 section2
15    132 section2
16    176 section2
17     52 section2
18    180 section2
```

Solution

Use the stack() function.

```
> stacked <- stack(unstacked)
> stacked
    values      ind
1      176 section1
2      176 section1
3       98 section1
4      118 section1
5      103 section1
6      190 section1
7      173 section1
8      184 section1
9      149 section1
10     120 section2
11     199 section2
12     159 section2
13     127 section2
14     141 section2
15     132 section2
16     176 section2
17      52 section2
18     180 section2
```

Similarly, you can unstack data by using the unstack() function:

```
> unstacked <- unstack(stacked)
> unstacked
  section1 section2
1     176      120
2     176      199
3      98      159
4     118      127
5     103      141
6     190      132
7     173      176
8     184       52
9     149      180
```

Here are some important points about the stack() and unstack() functions. First, you can only stack data on numeric variables. If there are more than two variables in the data frame, you must specify which variables to use, as in the following example. Because there were only two variables in the example, the values argument was not needed, but if there are more than two variables, you must specify the grouping variable (factor) to use for unstacking. The ~ (tilde) notation means the same as "by" to R. So we are telling R to unstack the data using the values "by" individual.

```
> unstacked2 <- unstack(stacked, values ~ ind)
> unstacked2
  section1 section2
1     176      120
2     176      199
3      98      159
4     118      127
5     103      141
6     190      132
7     173      176
8     184       52
9     149      180
```

If you have stacked data in which the number of values in each group differ, when you try to unstack that dataset, R cannot make a data frame and will output a list instead (see Recipe 3-3 for an example in which I created a list because two classes had different numbers of students.). You can still access the groups by using the $ notation that we have discussed when the data are in a list.

To illustrate, the complete data for the two sections is mismatched in length; that is, the classes are of different sizes. Here is the complete dataset. For each section, I prepared a statistics template in Microsoft Excel, and I could determine from the online classroom which students had downloaded and used the templates. I was interested in learning if the students who used the Excel template made better grades on the quiz.

```
  Score Section Used
1   176       1    0
2   176       1    0
3    98       1    0
4   118       1    0
5   103       1    0
6   190       1    1
7     0       1    0
```

8	173	1	0
9	0	1	0
10	184	1	0
11	149	1	0
12	0	1	0
13	136	1	0
14	171	1	0
15	174	1	0
16	155	1	0
17	154	1	1
19	199	2	1
20	159	2	1
21	127	2	0
22	141	2	0
23	0	2	0
24	132	2	0
25	176	2	1
26	0	2	0
27	52	2	0
28	180	2	1
29	120	2	0

When I unstack this set of data, I get lists rather than a data frame. The scores are coerced into character format.

```
unstacked <- unstack(quizGrades, Score ~ Section)
> unstacked
$`1`
 [1] "176" "176" "98"  "118" "103" "190" "0"   "173" "0"   "184" "149" "0"
[13] "136" "171" "174" "155" "154"

$`2`
 [1] "199" "159" "127" "141" "0"   "132" "176" "0"   "52"  "180" "120"

> typeof(unstacked)
[1] "list"
```

Finally, to the point of stacked and unstacked data, examine the output of the t.test() function using first the unstacked data and then the stacked data. As I mentioned, this function can handle both types of data. The t.test() function is more flexible than many others in R:

> t.test(unstacked$quiz1,unstacked$quiz2)

```
        Welch Two Sample t-test

data:  unstacked$quiz1 and unstacked$quiz2
t = 0.4774, df = 15.52, p-value = 0.6397
alternative hypothesis: true difference in means is not equal to 0
95 percent confidence interval:
 -31.06128  49.06128
sample estimates:
mean of x mean of y
 151.8889  142.8889
```

```
> t.test(stacked$score ~ stacked$section)

        Welch Two Sample t-test

data:  stacked$score by stacked$section
t = 0.4774, df = 15.52, p-value = 0.6397
alternative hypothesis: true difference in means is not equal to 0
95 percent confidence interval:
 -31.06128  49.06128
sample estimates:
mean in group 1 mean in group 2
      151.8889        142.8889
```

Working with Dates and Strings

In Chapter 5, you will learn how to work with dates and strings.

Recipe 5-1. Working with Dates and Times

Problem

When you import date and time data into R, these are not recognized automatically. To work with them, you must convert dates and times to the proper format.

Solution

The system's idea of the current date is returned by the Sys.Date() function. You can retrieve the current date with the time by using the Sys.time() function. Examine the following examples.

```
> Sys.time()
[1] "2014-06-03 13:26:35.06454 EDT"

> ## locale-specific version of date()
> format(Sys.time(), "%a %b %d %X %Y")
[1] "Tue Jun 03 1:26:35 PM 2014"
>
> Sys.Date()
[1] "2014-06-03"
>
> Sys.timezone()
[1] "America/New_York"
```

In R, the default format for dates is the four-digit year, followed by the month, and then the day. These can be separated by slashes or dashes and must be converted using the as.Date() function. R provides three date and date-time variable classes. These are Date, POSIXct, and POSIXlt. The current date and time are in PSOSIXct format by default, and this is generally the best alternative.

Here are some examples.

```
> as.Date("1952-5-30")
[1] "1952-05-30"
> as.Date("1952/10/28")
[1] "1952-10-28"
```

Many programs, such as Microsoft Excel, use the format month/day/year rather than year/month/day. To deal with this situation, you can create a format string using any of the following date codes (see Table 5-1).

Table 5-1. *R Format Codes for Various Date Values*

Code	Value
%d	Day of the month (decimal number)
%m	Month (decimal number)
%b	Month (abbreviated)
%B	Month (full name)
%y	Year (two digits)
%Y	Year (four digits)

For example, to convert the character string "6/1/2014" to a date in R, create the format string "%m/%d/%Y" to achieve the desired result.

```
> today <- as.Date("6/1/2014", format="%m/%d/%Y")
> class(today)
[1] "Date"
```

Date objects are stored internally as the number of days since January 1, 1970. Earlier dates are represented by negative numbers. You can convert a date object to the internal form by using the as.numeric() function. For example, statistician John Tukey's birthdate was 6/16/1915, while R. A. Fisher was born on 2/17/1890. We can use the weekdays() and months() functions to extract the desired components of a date:

```
StatBdays <- c(tukey = as.Date("1915-01-16"),fisher = as.Date("1890-02-17"))
> StatBdays
       tukey        fisher
"1915-01-16" "1890-02-17"
> weekdays(StatBdays)
     tukey      fisher
"Saturday"    "Monday"
> months(StatBdays)
     tukey      fisher
 "January" "February"
```

You can perform arithmetic with dates. For example, to determine the age in days of a person born on June 3, 2000, you could do the following. First, assign today's date to the variable today. Then, assign a date to June 3, 2000. Finally, subtract the dates as follows:

```
> today <- Sys.Date()
> today
[1] "2014-06-27"
> then <- as.Date("2000/6/3")
> then
[1] "2000-06-03"
> howLong <- today - then
> howLong
```

```
Time difference of 5137 days
> Sys.Date() - then
Time difference of 5137 days
```

You can make this a little more generic by using the system date instead of creating a date variable for the current day:

```
> Sys.Date() - then
Time difference of 5137 days
```

Here is a more meaningful use of date calculations. We have $1,000 to invest, and want to know how much money we will have on May 22, 2015, if we can earn simple interest of .05% per day. The `as.integer()` function converts the dates to integer format. I wrote a simple function to calculate the simple interest, and then supplied the appropriate arguments to it to find the answer.

```
> start <- as.integer(as.Date("2015/1/1"))
> stop <- as.integer(as.Date("2015/5/22"))
> t <- stop - start
>
> Return <- function(p = 1000, r = .0005, t = 365){
+ amount <- p * (1 + r * t)
+ return(amount)
+ }
> t
[1] 141
> Return(1000,.0005, 141)
[1] 1070.5
```

When you have times along with dates, the best class to use is most often POSIXct objects, as mentioned previously. The "ct" stands for *calendar time*. The POSIXlt object stands for *local time*. The POSIXct class returns the numeric value , whereas the POSIXlt class returns a list, as you can see from examining the following code. As the name implies, the `unclass()` function returns a copy of its argument with its class attribute removed.

```
> time1 <- as.POSIXct(Sys.Date())
> time1
[1] "2014-06-02 20:00:00 EDT"
> unclass(time1)
[1] 1401753600
> time2 <- as.POSIXlt(Sys.Date())
> unclass(time2)
$sec
[1] 0

$min
[1] 0

$hour
[1] 0

$mday
[1] 3
```

```
$mon
[1] 5

$year
[1] 114

$wday
[1] 2

$yday
[1] 153

$isdst
[1] 0

attr(,"tzone")
[1] "UTC"
> time1
[1] "2014-06-02 20:00:00 EDT"
> class(time1)
[1] "POSIXct" "POSIXt"
> mode(time1)
[1] "numeric"
> time2
[1] "2014-06-03 UTC"
> class(time2)
[1] "POSIXlt" "POSIXt"
> mode(time2)
[1] "list"
```

Recipe 5-2. Working with Character Strings
Problem

Visualize an iceberg. When we think of data, we typically think of something like a data frame in R, an Excel spreadsheet, or some other database with a fixed structure. As you have seen, data frames can include both string data (character) as numbers, but in data frames, these are limited to a fixed structure and represent factors or nominal variables. The visible part of the iceberg is about 20%, containing "data" as we commonly conceptualize it. The 80% below the surface contains a dizzying array of "stuff," and much of that stuff is very useful, even vital, to us on a daily basis, both at a personal and at a business level. The stuff includes, among other things, video and audio files, images, texts of all kinds, PDF files, PowerPoint files, e-mail, notes, and Word documents.

We have a digital universe that is growing exponentially. According to EMC's seventh digital universe study conducted by the market research company IDC, the size of the digital universe increases 40% per year.

In the year 2005, there were "only" 132 exabytes of data. An *exabyte* is 2.5×10^{18} bytes. The Internet of Things (IoT) is predicted to account for approximately 10% of the digital universe by 2020, which itself will contain nearly as many digital bits as the number of stars in the "real" universe.

With this much information "out there," and the majority of it not numbers, but instead narratives, pictures, and sounds, text mining has become increasingly important. Although perhaps not as proficient as other scripting languages in this regard, R is still quite capable of working with string data. We will discuss creating strings first, and then we will discuss various options for working with string data.

Solution

The class of a string object is a *character*. Strings must be enclosed in either single or double quotes. You can insert single quotes into a string enclosed in double quotes, and vice versa, but you cannot insert the same kind of quote. In order for R to recognize the quote, you have to escape it with a slash. The character() function is used to create vector objects of the character type.

In this solution, you'll learn how to create character strings and how to find patterns and matches in strings. You will also learn how to use the stringr() function to make working with strings more effective and more systematic.

Creating Character Strings

You can create character strings in a couple of different ways. You have already seen the use of the c() function. You can also create an empty character vector and then fill in the elements separately. Here are a couple of examples.

```
> example <- character(5)
> example
[1] "" "" "" "" ""
> example[1] <- "a"
> example[2] <- "b"
> example[3] <- "c"
> example[4] <- "d"
> example[5] <- "e"
> example
[1] "a" "b" "c" "d" "e"
> example2 <- c("f","g","h","i","j")
> example2
[1] "f" "g" "h" "i" "j"
> c(example, example2)
 [1] "a" "b" "c" "d" "e" "f" "g" "h" "i" "j"
```

Another important function for dealing with character data is paste(). This function takes any number of arguments, coerces them to character type if they are not already in that format, and then pastes, or *concatenates*, the arguments into one or more character strings. Examine the following code segments.

```
> PieLife <- paste("The life of", pi,"is sweet")
> PieLife
[1] "The life of 3.14159265358979 is sweet"
```

The default is to use a space as the separator, but you can specify other separators by declaring the type you want. For example, you can use a comma followed by a space.

```
> MyPieLife <- paste("Today", Sys.Date(), "I did not eat pie.", sep = ", ")
> MyPieLife
[1] "Today, 2014-06-06, I did not eat pie."
> typeof(MyPieLife)
[1] "character"
```

You can also use the cat() function to concatenate output, but as the following example shows, one cannot save the output from the cat() function to a variable. You will find the cat() and print() functions to be very useful when you write your own custom functions in R. Note the "escaped" character "\n" to tell R to go to the next line. Without the "\n", R would keep the command prompt on the same line with the output. Observe that the attempt to create a variable called MyPieLife with the paste() function was successful, but the same thing is not true for the cat() function. The "variable" we created is nonexistent.

```
> MyPieLife <- cat("Today, ", as.character(Sys.Date()),", I did not eat pie.","\n")
Today,  2014-06-06 , I did not eat pie.
>
> typeof(MyPieLife)
[1] "NULL"
```

Pasting has the same recycling property as vectors do. If you paste objects of different lengths, the shorter length will be recycled, as you can see in the following example.

```
> paste("X",1:10,sep = ".")
 [1] "X.1"  "X.2"  "X.3"  "X.4"  "X.5"  "X.6"  "X.7"  "X.8"  "X.9"  "X.10"
```

In addition to concatenating character values with the paste() function, you can also use the sprintf() function. This function allows us the opportunity to control output by specifying the format of the objects being printed. For example:

```
> sprintf("%s was born in %d", "Tukey", 1915)
[1] "Tukey was born in 1915"
```

Finding Patterns and Matches in Strings

The substr() function can be used to extract a substring, and the sub() function can be used to replace the first occurrence of a word or substring match. The gsub() function replaces all matches. The grep() function searches for matches to a pattern within the elements of a character vector. See the following examples for the use of the sub() and gsub() functions.

```
> TukeySaid <- "An approximate answer to the right problem is worth a good
+ deal more than the exact answer to an approximate problem."
> substr(TukeySaid,start = 3, stop = 14)
[1] " approximate"
> sub("answer", "solution", TukeySaid)
[1] "An approximate solution to the right problem is worth a good\ndeal more than the exact answer
to an approximate problem."
> gsub("answer", "solution", TukeySaid)
[1] "An approximate solution to the right problem is worth a good\ndeal more than the exact solution
to an approximate problem."
>
```

The grep() function can locate matches in character vectors. For example, the state.name dataset that ships with R lists the names of the 50 United States:

```
state.name
 [1] "Alabama"        "Alaska"         "Arizona"        "Arkansas"
 [5] "California"     "Colorado"       "Connecticut"    "Delaware"
 [9] "Florida"        "Georgia"        "Hawaii"         "Idaho"
[13] "Illinois"       "Indiana"        "Iowa"           "Kansas"
[17] "Kentucky"       "Louisiana"      "Maine"          "Maryland"
[21] "Massachusetts"  "Michigan"       "Minnesota"      "Mississippi"
[25] "Missouri"       "Montana"        "Nebraska"       "Nevada"
[29] "New Hampshire"  "New Jersey"     "New Mexico"     "New York"
[33] "North Carolina" "North Dakota"   "Ohio"           "Oklahoma"
[37] "Oregon"         "Pennsylvania"   "Rhode Island"   "South Carolina"
```

```
[41] "South Dakota"   "Tennessee"    "Texas"        "Utah"
[45] "Vermont"        "Virginia"     "Washington"   "West Virginia"
[49] "Wisconsin"      "Wyoming"
```

Let us find the states with the word "New" in their names. Without the argument value = TRUE, the grep() function returns the index numbers of these states rather than their names. See the following:

```
> grep(state.name, pattern = "New")
[1] 29 30 31 32
> grep(state.name, pattern = "New", value = TRUE)
[1] "New Hampshire" "New Jersey"    "New Mexico"    "New York"
```

Now, find the state or states with the longest name(s). Use the nchar() function for this purpose. We see that two of the states have 14-character names. We can then determine the names of those two states.

```
> nchar(state.name)
 [1]  7  6  7  8 10  8 11  8  7  7  6  5  8  7  4  6  8  9  5  8 13  8  9 11  8
[26]  7  8  6 13 10 10  8 14 12  4  8  6 12 12 14 12  9  5  4  7  8 10 13  9  7
> longest <- nchar(state.name)
> state.name[which(longest == max(longest))]
[1] "North Carolina" "South Carolina"
```

Using the stringr Package

The stringr package written by Hadley Wickham overcomes some of the limitations of the base version of R when it comes to string manipulations. According to Wickham, the stringr package is a "set of simple wrappers that make R's string functions more consistent, simpler, and easier to use." Install the package by using the install.packages() function.

You can load stringr into your current R session with library() or require(). To see the list of the functions available in stringr, use the command library(help = stringr).

```
> install.packages("stringr")
> library(stringr)
> library(help = stringr)
```

The stringr package has all the functionality of the string functions we have used previously, but has the advantage that it works with missing data in a more appropriate way, as demonstrated next. The stringr package also has functionality that is not available in base R, such as the ability to duplicate characters. All the functions start with "str_" followed by a term that is descriptive of the task the function performs.

The str_length() and str_c() Functions

In the base R string functions, NA is treated as a two-character string, rather than as missing data. To illustrate, the nchar() function counts the characters in "NA" and reports it as a two-character string, while the str_length() function in stringr recognizes the missing value as such:

```
myName <- c("Larry",NA,"Pace")
> nchar(myName)
[1] 5 2 4
> str_length(myName)
[1]  5 NA  4
```

The str_length() function also converts factors to characters, something of which nchar() is not capable.

```
> sexFactor <- factor(c(0,0,0,0,1,1,1,1,0,1,1,0,0,1), labels = c("female","male"))
> sexFactor
 [1] female female female female male   male   male   male   female male
[11] male   female female male
Levels: female male
> nchar(sexFactor)
Error in nchar(sexFactor) : 'nchar()' requires a character vector
> str_length(sexFactor)
 [1] 6 6 6 6 4 4 4 4 6 4 4 6 6 4
```

The str_c() function is a substitute for paste(), but uses the empty string "" as the default separator instead of using the whitespace, as paste() does.

```
> str_c("Statistics","is","the","grammar","of","science.","Karl Pearson")
[1] "Statisticsisthegrammarofscience.Karl Pearson"
```

You can change the separator by using the sep argument, as follows:

```
> str_c("Statistics","is","the","grammar","of","science.","Karl Pearson", sep = " ")
[1] "Statistics is the grammar of science. Karl Pearson"
```

The str_sub() Function

The str_sub() function extracts substrings from character vectors. The user supplies three arguments: the string vector, the start value, and the end value. The function has the ability to work with negative indexes, which cause the function to work backward from the last character in a string element.

```
> pearsonSays <-str_c("Statistics","is","the","grammar","of","science.","Karl Pearson", sep = " ")
> pearsonSays
[1] "Statistics is the grammar of science. Karl Pearson"
> str_sub(pearsonSays, start = 1, end = 10)
[1] "Statistics"
> str_sub(pearsonSays, start = -7, end = -1)
[1] "Pearson"
```

You can also use the str_sub() function to replace substrings, as in the following example.

```
> str_sub(pearsonSays, 39, 50) <- "Ronald Fisher"
> pearsonSays
[1] "Statistics is the grammar of science. Ronald Fisher"
```

The str_dup() Function

R provides no specific function for duplicating string characters, but the str_dup() function in stringr allows that operation. The str_dup() function duplicates and then concatenates strings within a character vector. You can specify the particular string as well as the number of times the string is to be duplicated. See the following:

```
> SantaSays <- str_dup("Ho", 3)
> SantaSays
[1] "HoHoHo"
```

```
> MrsSantaSays <- c(str_dup("Merry", 1:3),"Christmas")
> MrsSantaSays
[1] "Merry"           "MerryMerry"      "MerryMerryMerry" "Christmas"
```

Padding, Wrapping, and Trimming Strings

Padding involves taking a string and adding leading or trailing characters (or both) to achieve a specified width. The str_pad() function accomplishes this. The default is the use of a space (pad = " "). The side argument takes the options "left", "right", and "both" to achieve left, right, and centered alignment. Here are some examples.

```
> str_pad("Tukey", width = 10)
[1] "     Tukey"
> str_pad("Tukey", width = 10, side = "right")
[1] "Tukey     "
> str_pad("Tukey", width = 10, side = "both")
[1] "  Tukey   "
> str_pad("Tukey", width = 10, pad = "#")
[1] "#####Tukey"
```

The str_wrap() function can wrap a string to form paragraphs. For example, consider the following quote from R. A. Fisher:

```
fisherSays <- c(
        "If ... we choose a group of social",
        "phenomena with no antecedent knowledge",
        "of the causation or absence of causation",
        "among them, then the calculation of",
        "correlation coefficients, total or partial,",
        "will not advance us a step toward evaluating",
        "the importance of the causes at work.",
        "R. A. Fisher"
        )
```

To display this quote as a single paragraph, we must paste the elements together as follows. The collapse argument tells R to "unconcatenate" the individual lines and create a single string vector.

```
fisherSays <- paste(fisherSays, collapse = " ")
```

We can control the width of the lines, as well as indentation. The default arguments for indent and exdent are 0. Here is an example.

```
> cat(str_wrap(fisherSays, width = 30, indent = 2), "\n")
  If ... we choose a group of
social phenomena with no
antecedent knowledge of the
causation or absence of
causation among them, then
the calculation of
correlation coefficients,
total or partial, will not
```

```
advance us a step toward
evaluating the importance of
the causes at work. R. A.
Fisher
```

We can trim strings using the str_trim( ) function. In string processing, we often parse a text into individual words. The words usually wind up having whitespaces (blank space) on either end. If that is the situation, use str_trim( ) to remove the whitespaces.

```
> textToTrim <- c("There", "   are","many   "," extra ", "whitespaces")
> textToTrim
[1] "There"        "   are"        "many   "       " extra "       "whitespaces"
> str_trim(textToTrim, side = "both")
[1] "There"        "are"          "many"          "extra"         "whitespaces"
```

Extracting Words

The word( ) function extracts words from a sentence. You pass the function a string along with the starting position of the first word to extract. The end position is that of the last word to extract. By default, a single space is used as the separator between words. Let's use the Fisher quote and extract different words. We extract the first, the second, and the last words of each string.

```
> fisherSays <- c(
+ "If ... we choose a group of social",
+ "phenomena with no antecedent knowledge",
+ "of the causation or absence of causation",
+ "among them, then the calculation of",
+ "correlation coefficients, total or partial,",
+ "will not advance us a step toward evaluating",
+ "the importance of the causes at work.",
+ "R. A. Fisher")
> word(fisherSays, 1)
[1] "If"           "phenomena"    "of"            "among"         "correlation"
[6] "will"         "the"          "R."
> word(fisherSays, 2)
[1] "..."          "with"          "the"           "them,"
[5] "coefficients," "not"          "importance"    "A."
> word(fisherSays, -1)
[1] "social"       "knowledge"    "causation"     "of"            "partial,"
[6] "evaluating"   "work."        "Fisher"
```

CHAPTER 6

Working with Tables

R provides a variety of ways to work with data arranged in tabular form. Previous chapters covered vectors, matrices, lists, factors, and data frames. This chapter expands on this and covers how to work with tables. We will start with simple frequency tables in which we summarize the frequencies of observations in each level of a categorical variable. We will then consider two-dimensional tables and higher-order tables. You will learn how to create, display, and analyze tabular data. We will limit ourselves to categorical data, but we will discuss frequency distributions for scale (interval and ratio) data in Chapter 7.

The table() function returns a contingency table, which is an object of class table and is an array of integer values. The integer values can be arranged in multiple rows and columns, just as a vector can be made into a matrix. As an example, the HairEyeColor data included with R are in the form of a three-way table, as the following code demonstrates:

```
> data(HairEyeColor)
> HairEyeColor
, , Sex = Male

       Eye
Hair    Brown Blue Hazel Green
Black      32   11    10     3
Brown      53   50    25    15
Red        10   10     7     7
Blond       3   30     5     8

, , Sex = Female

       Eye
Hair    Brown Blue Hazel Green
Black      36    9     5     2
Brown      66   34    29    14
Red        16    7     7     7
Blond       4   64     5     8

> class(HairEyeColor)
[1] "table"
```

Recipe 6-1. Working with One-Way Tables
Problem

Researchers often collect and enter data in haphazard order. We can sort the data to make more sense of it, but a frequency table makes even more sense, and is often one of the first things we will do when we have numbers that are not in any particular order. If we count the frequencies of each raw data value, we have created a simple frequency distribution, also known as a *frequency table*. If there is only one categorical variable, the table() function will return a one-way table. For those unfamiliar with this method of describing tables, a one-way table has only one row or column of data.

Solution

The following data came from the larger plagiarism study I conducted. For 21 students, the major source of the material they plagiarized was determined though Turnitin originality reports as being the Internet, other student papers, or publications such as articles in journals. The data were not in any particular order. The table() function makes it easy to determine the Internet and other student papers were the most popular sources of plagiarized material. Incidentally, many students tell me when confronted that they did not know it was wrong to paste material from web sites into their papers without attribution of the source. One even told me she had done this throughout her entire master's degree program and no one had ever corrected her.

```
> Source
 [1] "Internet"      "Internet"      "Publication"   "StudentPapers"
 [5] "Internet"      "Internet"      "Internet"      "Internet"
 [9] "StudentPapers" "Publication"   "StudentPapers" "Internet"
[13] "StudentPapers" "StudentPapers" "StudentPapers" "Internet"
[17] "StudentPapers" "StudentPapers" "Internet"      "Internet"
[21] "StudentPapers"
> table(Source)
Source
    Internet   Publication StudentPapers
          10             2             9
```

You can also make a table from summary data if you have such data already collected. For example, the following data are from a class exercise in which I have students count the colors of the paint finishes on the first 100 cars and trucks they find in a parking lot or passing through an intersection. The table can be made directly by use of the as.table() function. The names will default to letters of the alphabet, but you can overwrite those with more meaningful names. To conserve space, we will use the first letter or the first two letters of the color names.

```
> carColors <- as.table(c(19,16,17,16,6,7,4,6,6,3))
> carColors
 A  B  C  D  E  F  G  H  I  J
19 16 17 16  6  7  4  6  6  3
> row.names(carColors) <- c("W","B","S","G","R","Bl","Br","Y","G","O")
> carColors
 W  B  S  G  R Bl Br  Y  G  O
19 16 17 16  6  7  4  6  6  3
```

When presenting one-way frequency distributions visually, you should use either a pie chart or a bar chart. The categories are separate, and there is no underlying continuum, so the bars in a bar chart should not touch. To make the graphics more meaningful, you can color the bars or pie slices to match the colors represented in the table.

Chapter 8 covers data visualization in more depth. For now, let's just create a bar plot of the car color data and give the bars the appropriate colors (see Figure 6-1). You can find the named colors that R supports by doing a quick Internet search. In the following code, see that the *x*- and *y*-axis labels can be specified by use of the xlab and ylab arguments.

```
> myColors <- c("white","black","gray","gray50","red","blue","brown","yellow","green","khaki")
> barplot(carColors, col = myColors, xlab = "Color", ylab = "Frequency", main = "Automotive + Paint
Color Preferences")
```

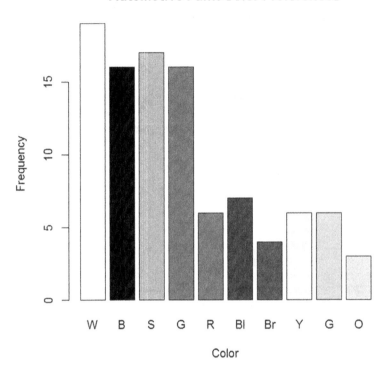

Figure 6-1. *Automotive paint color preferences*

We can also determine whether the frequencies in a table are distributed on the basis of chance, or if they conform to some other expectation. The most common way of doing this for one-way tables is the chi-square test we will discuss in greater detail in Chapter 10.

Recipe 6-2. Working with Two-Way Tables
Problem

The analysis of cross-classified categorical data occupies a prominent place in statistics and data analysis. A wide variety of biological, business, and social science data take the form of cross-classified contingency tables. Although a commonly used practice in the analysis of contingency tables is the chi-square test, modern alternatives are available, in particular log-linear models. In Recipe 6-2, you will learn how to create and display two-way tables.

Solution

Until the recent past, the statistical and computational techniques for analyzing cross-classified data were limited, and the typical approach to analyzing higher order tables was to do a series of analyses of two-dimensional marginal totals. The chi-square test has stood the test of time, but the recent development of log-linear models makes it possible to overcome the limitations of the two-dimensional approach when there are tables with more than two categories.

As with one-way tables, you can create a contingency table from raw data, or you can create a table from summary data with the as.table() function. I will illustrate both options. Let us return to the mouse weight dataset discussed in Chapter 2. The strain of the mouse and the sex of the mouse are both categorical data. We will use the table() function to create a two-way table of sex by strain. Then, we will produce a nice-looking clustered barplot of the data for visualizing the distribution of sex by strain. The strains have been labeled A, B, C, and D.

```
> counts <- table(mouseWeights$sex,mouseWeights$strain)
> counts

    A   B   C   D
 F  62 178 137  41
 M  36 198 131  61
> barplot(counts, col = c("gray","antiquewhite"), ylab = "Count",xlab = "Strain",
+ main = "Mouse Strain by Sex", legend = rownames(counts), beside=T)
```

The clustered or "side-by-side" bar graph is shown in Figure 6-2.

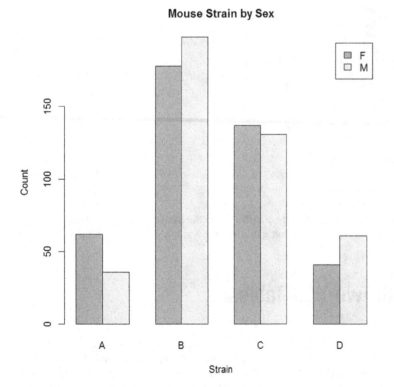

Figure 6-2. *Clustered bar graph of mouse sex by strain*

Now, let's create a two-way table from summary data. We'll return to the GSS data and make a two-way table that compares the job satisfaction scores from 1 to 7 for males and females (labeled 1 and 2, respectively). Here is the table:

```
    1    2    3    4    5   6   7
1  66  144  116  29  25  10  5
2  56  120  144  25  15   3  4
```

We will use the as.table() function once again to create the two-way contingency table. To make the table, I first created a vector, then turned the vector into a matrix, and finally made the matrix into a table. As before, the default row and column names are successive letters of the alphabet, but it is easy to change those to the correct labels, in this case, numbers.

```
> newTable <- c(66, 144, 116, 29, 25, 10, 5, 56, 120, 144, 25, 15, 3, 4)
> newTable <- matrix(newTable, ncol = 7, byrow = T)
> newTable
      [,1] [,2] [,3] [,4] [,5] [,6] [,7]
[1,]   66  144  116   29   25   10    5
[2,]   56  120  144   25   15    3    4
> newTable <- as.table(newTable)
> newTable
    A   B   C   D   E   F   G
A  66 144 116  29  25  10   5
B  56 120 144  25  15   3   4
> rownames(newTable) <- 1:2
> colnames(newTable) <- 1:7
> newTable
    1   2   3   4   5   6   7
1  66 144 116  29  25  10   5
2  56 120 144  25  15   3   4
```

As before, we can make a barplot to show the distribution of job satisfaction by sex (see Figure 6-3).

```
> barplot(newTable, beside = T, xlab = "Job Satisfaction", ylab = "Frequency",
+ legend = rownames(newTable))
```

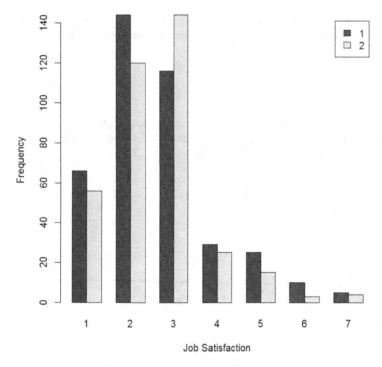

Figure 6-3. Job satisfaction scores by sex (source: GSS)

Recipe 6-3. Analyzing One- and Two-Way Tables
Problem

A very common problem is determining whether the observed frequencies in one- or two-way tables differ significantly from expectation. We determine an expected frequency based on some null hypothesis, and then compare the observed and expected frequencies to determine whether the observed frequencies are due to chance (sampling error) or due to significant departures from the expected frequencies. The most frequently used tool for this kind of tabular analysis is the chi-square test.

The chi-square test comes in two varieties. When we analyze a one-way table, we use a chi-square test of goodness of fit. Two-way tables are analyzed with a chi-square test of independence (also known as a *test of association*). For the goodness-of-fit test, the expected frequencies may be equal or unequal, depending on the null hypothesis. The test of independence, on the other hand, uses the observed frequencies to calculate the expected frequencies. I will explain and illustrate both tests.

Solution

The German geodist Friedrich Robert Helmert discovered the chi-square distribution in 1876. Helmert found that the sample variance for a normal distribution followed a chi-square distribution. Helmert's discovery was touted in his own books, as well as other German texts, but was not known in English. The chi-square distribution was independently rediscovered in 1900 when Karl Pearson developed his goodness-of-fit test. R. A. Fisher and William S. Gosset ("Student") later made the connection to the sample variance.

The chi-square distribution was applied to tests of goodness of fit by Pearson. R—along with many statistical packages, such as SPSS—labels the test "Pearson's Chi-Square." The chi-square distribution has a single parameter, known as the *degrees of freedom*. Unlike other distributions, the chi-square distribution's degrees of freedom are based on the number of categories rather than the sample size. Pearson had difficulty applying the concept of degrees of freedom correctly to his own chi-square tests, and was subsequently corrected by Fisher on this matter. The degrees of freedom are the number of categories minus 1. The total of the observations, N, and the total of the number of categories, k, are both constrained. The N objects must be placed in the k categories in such a way that the probability totals 1, and each object fits into one (and only one) of the categories. These are the conditions of mutual exclusivity and exhaustiveness. Interestingly, from a mathematical standpoint, the expected value of chi-square when the null hypothesis is true is the degrees of freedom. As the value of chi-square increases, it becomes less probable that the observed values came from the distribution specified by the null hypothesis.

We will do a chi-square goodness-of-fit test for the car color preference data. Assuming the 10 colors would be distributed evenly if each color was equally preferred, we would expect each color to occur 10 times in the sample of 100 cars. To calculate chi-square, we subtract the expected frequency from the observed frequency for each category, square this deviation score, divide the squared deviation by the expected frequency, and sum these values across categories. As explained earlier, the degrees of freedom for this particular test would be $10 - 1 = 9$. Clearly, the colors are not equally popular, as is shown by the p value, which is substantially lower than .05, the customary alpha level.

```
> carColors
 W  B  S  G  R Bl Br  Y  G  O
19 16 17 16  6  7  4  6  6  3
> chisq.test(carColors)

        Chi-squared test for given probabilities

data: carColors
X-squared = 34.4, df = 9, p-value = 7.599e-05
```

For the chi-square test of independence, the expected frequencies are calculated by multiplying the marginal (row and column) totals for each cell, and then dividing that product by the overall sample size. The resulting values are the frequencies that would be expected if there were no association between the two categories. It is customary to identify the two-way contingency table as $r \times c$ (row-by-column), and the degrees of freedom for the chi-square test of independence are calculated as the number of rows minus 1 times the number of columns minus 1, or $(r - 1)(c - 1)$.

Let us determine if the sexes of the mice in our dataset are equally distributed across the four strains. Once again, the p value lower than .05 makes it clear that there are unequal numbers of male and female mice across the strains.

```
> chisq.test(mouseWeights$sex, mouseWeights$strain)

        Pearson's Chi-squared test

data: mouseWeights$sex and mouseWeights$strain
X-squared = 11.9429, df = 3, p-value = 0.007581
```

The expected frequency for each cell, as I mentioned, is calculated by multiplying the marginal row total and the marginal column total for that cell, and dividing the product by the overall sample size. Here is the table with the marginal totals added:

```
> newTable
            A   B   C   D rowMargins
F          62 178 137  41        418
M          36 198 131  61        426
colMargins 98 376 268 102        844
```

73

As an example, the expected number of female mice for strain A would be found by multiplying 418 by 98, and dividing the product, 40964, by the number of mice, 844. The resulting value of 48.54 is the expected frequency if the sex and strain of the mice were not associated. With a little matrix algebra, it is easy to calculate all the expected frequencies at once:

```
> rowMargins <- matrix(c(418,426)rowMargins)
> rowMargins
      [,1]
[1,]   418
[2,]   426
> colMargins <- matrix(c(93,376,268,102))
> colMargins
      [,1]
[1,]    93
[2,]   376
[3,]   268
[4,]   102
> colMargins <- t(colMargins)
> expected <- (rowMargins %*% colMargins)/844
> expected
          [,1]     [,2]      [,3]     [,4]
[1,] 48.53555 186.218 132.7299 50.51659
[2,] 49.46445 189.782 135.2701 51.48341
```

Recipe 6-4. Working with Higher-Order Tables
Problem

As mentioned at the beginning of this chapter, the traditional way to analyze three-way or higher tables was to separate the table into a series of two-way tables and then perform chi-square tests of independence for each two-way table. Log-linear modeling of contingency tables has essentially supplanted the use of chi-square analysis for three-way and higher tables. I will illustrate using the aforementioned HairEyeColor table that ships with R. Here are the numbers:

```
> HairEyeColor
, , Sex = Male

       Eye
Hair    Brown Blue Hazel Green
Black      32   11    10     3
Brown      53   50    25    15
Red        10   10     7     7
Blond       3   30     5     8

, , Sex = Female

       Eye
Hair    Brown Blue Hazel Green
Black      36    9     5     2
Brown      66   34    29    14
Red        16    7     7     7
Blond       4   64     5     8
```

The customary way of handling this sort of data would be to do a chi-square test of independence for each sex to determine if there is an association between hair color and eye color. This, of course, omits the consideration of the possible interactions between sex and the other two categorical variables.

Solution

The log-linear model provides the ability to analyze the three-way table as a whole. Log-linear models range from the simplest model, in which the expected frequencies are all equal, to the most complex (saturated model), in which every component is included.

To do a log-linear analysis, we load the data and the MASS package. The log-linear analysis produces both a Pearson's chi-square and a likelihood-ratio chi-square. We are testing the same null hypothesis that we do in a two-way chi-square, namely that the three or more categorical variables are independent.

We build a formula to indicate the categorical variables in our analysis, and we use different signs to indicate how we want to combine these variables. The tilde (~) indicates the beginning of the formula. The plus signs indicate an additive model; that is, one in which we are not interested in examining interactions. By contrast, a multiplicative model uses asterisks (*) to indicate a model with interactions. Here is the additive model for the hair, sex, and eye color data:

```
> data(HairEyeColor)
> library(MASS)
> dimnames(HairEyeColor)
$Hair
[1] "Black" "Brown" "Red" "Blond"

$Eye
[1] "Brown" "Blue" "Hazel" "Green"

$Sex
[1] "Male" "Female"

> indep <- loglm(~Hair + Eye + Sex, data = HairEyeColor)
> summary(indep)
Formula:
~Hair + Eye + Sex
attr(,"variables")
list(Hair, Eye, Sex)
attr(,"factors")
     Hair Eye Sex
Hair    1   0   0
Eye     0   1   0
Sex     0   0   1
attr(,"term.labels")
[1] "Hair" "Eye" "Sex"
attr(,"order")
[1] 1 1 1
attr(,"intercept")
[1] 1
attr(,"response")
[1] 0
attr(,".Environment")
<environment: R_GlobalEnv>
```

```
Statistics:
                         X^2  df  P(>X^2)
Likelihood Ratio  166.3001  24        0
Pearson           164.9247  24        0
```

With *p*-values of approximately zero, both hypothesis tests show that the model departs significantly from independence.

A fully-saturated model would include the main effects along with all the interactions, and would not provide useful information, as the saturated model uses all the components and produces a likelihood-ratio statistic of zero:

```
> summary(indep2)
Formula:
~Hair * Eye * Sex
attr(,"variables")
list(Hair, Eye, Sex)
attr(,"factors")
     Hair Eye Sex Hair:Eye Hair:Sex Eye:Sex Hair:Eye:Sex
Hair    1   0   0        1        1       0            1
Eye     0   1   0        1        0       1            1
Sex     0   0   1        0        1       1            1
attr(,"term.labels")
[1] "Hair"         "Eye"         "Sex"         "Hair:Eye"    "Hair:Sex"
[6] "Eye:Sex"      "Hair:Eye:Sex"
attr(,"order")
[1] 1 1 1 2 2 2 3
attr(,"intercept")
[1] 1
attr(,"response")
[1] 0
attr(,".Environment")
<environment: R_GlobalEnv>

Statistics:
                   X^2  df  P(>X^2)
Likelihood Ratio    0   0        1
Pearson             0   0        1
```

We use an iterative process to find the least complex model that best explains the associations among the variables. This can be a hit-or-miss proposition. For example, the model ~ Eye + Hair * Sex stipulates that hair color and sex are dependent (associated), while eye color is independent of both hair color and sex.

```
> indep3 <- loglm(Eye ~ Hair * Sex, data = HairEyeColor)
> summary(indep3)
Formula:
Eye ~ Hair * Sex
attr(,"variables")
list(Eye, Hair, Sex)
attr(,"factors")
```

```
      Hair Sex Hair:Sex
Eye    0   0        0
Hair   1   0        1
Sex    0   1        1
attr(,"term.labels")
[1] "Hair"     "Sex"        "Hair:Sex"
attr(,"order")
[1] 1 1 2
attr(,"intercept")
[1] 1
attr(,"response")
[1] 1
attr(,".Environment")
<environment: R_GlobalEnv>

Statistics:
                    X^2  df  P(>X^2)
Likelihood Ratio 299.4790  24      0
Pearson          334.0596  24      0
```

You can see that this model is more effective than the initial additive one. We can use backward elimination, in which the highest-order interactions are successively removed from the saturated model until the reduced model no longer accurately describes the data. Although this process is not built into the loglm function, it is possible to build the saturated model and then use the step() function to with the direction set to backward:

```
> saturated <- loglm( ~ Hair * Eye * Sex, data = HairEyeColor)
> step(saturated, direction = "backward")
Start: AIC=64
~Hair * Eye * Sex

                  Df    AIC
- Hair:Eye:Sex  9 52.761
<none>             64.000

Step: AIC=52.76
~Hair + Eye + Sex + Hair:Eye + Hair:Sex + Eye:Sex

            Df     AIC
- Eye:Sex   3   51.764
<none>          52.761
- Hair:Sex  3   58.327
- Hair:Eye  9 184.678

Step: AIC=51.76
~Hair + Eye + Sex + Hair:Eye + Hair:Sex

            Df     AIC
<none>          51.764
- Hair:Sex  3   53.857
- Hair:Eye  9 180.207
```

```
Call:
loglm(formula = ~Hair + Eye + Sex + Hair:Eye + Hair:Sex, data = HairEyeColor,
    evaluate = FALSE)

Statistics:
                        X^2  df    P(>X^2)
Likelihood Ratio 11.76372   12  0.4648372
Pearson          11.77059   12  0.4642751
```

AIC is the Akaike Information Criterion. Smaller AIC values are better. In two steps, R has first eliminated the three-way interaction term, and then the two-way interaction between eye color and sex. Our most parsimonious model includes hair color, eye color, sex, and the interactions of hair and eye color and hair color and sex.

■ ■ ■

## CHAPTER 7

Summarizing and Describing Data

In this chapter, you learn how to summarize and describe data measured at the interval or ratio level. We will lump interval and ratio measures together and call the combination "scale" data because these types of measurement have equal intervals between successive data values.

Chapter 6 showed how to summarize categorical data using the table() function. This function works in the same fashion for scale data to produce simple frequency tables. With small datasets, simple frequency tables are fine, but with larger datasets, we need a way to group the data by intervals to keep the frequency table from being too long to be of any use.

In addition to simple and grouped frequency distributions, you will learn how to calculate the common descriptive statistics for scale data, including measures of central tendency, variability, skewness, and kurtosis. We will quickly outgrow the statistical prowess of the base R distribution and find that several packages will help us (and keep us from having to write custom functions). We will explore the usefulness of the fBasics and prettyR packages, in particular. The chapter ends with a description of how to calculate various quantiles for a data distribution.

Recipe 7-1. Creating Simple Frequency Distributions
Problem

We are awash in a sea of data. The volume of data is doubling approximately every 18 months. Raw data are not particularly helpful because it is hard to detect patterns. In addition to summarizing categorical data, tables are also very useful for summarizing the frequencies of interval and ratio data.

Solution

Tables help to make order from chaos. As mentioned, we use the table() function to create a simple frequency distribution. The simple frequency distribution tells us a great deal about the shape of the data, whether there is a modal value (or even multiple modes), and the range of data values. As mentioned, simple frequency distributions are limited to small datasets.

The ages in years of the recipients of the Oscar for Best Actress and Best Actor since the beginning of the Academy Awards in 1928 are in the data frame called oscars. I created the data frame from information available on Wikipedia. Interested readers can retrieve a copy of this data frame from the companion web site for this book.

```
> head(oscars)
    award              name
1 Actress      Janet Gaynor
2 Actress      Mary Pickford
3 Actress      Norma Shearer
4 Actress     Marie Dressler
5 Actress        Helen Hayes
6 Actress Katharine Hepburn
```

```
                                                         movie years days
1 Seventh Heaven, Street Angel, and Sunrise: A Song of Two Humans    22  222
2                                                       Coquette    37  360
3                                                    The Divorcee    28   87
4                                                     Min and Bill    63    1
5                                          The Sin of Madelon Claudet    32   39
6                                                   Morning Glory    26  308
> tail(oscars)
      award                name            movie years days
169 Actor          Sean Penn            Milk    48  189
170 Actor        Jeff Bridges    Crazy Heart    60   93
171 Actor         Colin Firth  The King's Speech    50  170
172 Actor       Jean Dujardin     The Artist    39  252
173 Actor    Daniel Day-Lewis        Lincoln    55  301
174 Actor Matthew McConaughey Dallas Buyers Club    44  118
```

We can construct a frequency table of the ages in years of the winners of Best Actress as follows:

```
> actress <- subset(oscars, award == "Actress")
> table(actress$years)

21 22 24 25 26 27 28 29 30 31 32 33 34 35 36 37 38 39 41 42 44 45 48 49 54 60
 1  2  2  4  5  4  4  8  3  3  4  6  3  5  2  2  4  2  6  2  1  2  1  2  1  1
61 62 63 74 80
 3  1  1  1  1
```

Let's do the same thing for the actors:

```
> actor <- subset(oscars, award == "Actor")
> table(actor$years)

29 30 31 32 34 35 36 37 38 39 40 41 42 43 44 45 46 47 48 49 50 51 52 53 54 55
 1  2  1  3  3  2  4  5  5  5  3  4  5  5  2  5  1  4  3  4  2  2  2  2  1  1
56 57 60 62 76
 1  2  3  3  1
```

With 87 people in each subset, we are beginning to see the limits of the simple frequency distribution. With this many data points, it seems a grouped frequency distribution would be more helpful (see Recipe 7-2).

Recipe 7-2. Creating Grouped Frequency Distributions

Problem

Simple frequency distributions are not appropriate for large datasets. The "granularity" of the data at the level of the individual values presents a "forest versus trees" dilemma (see Recipe 7-1 for tables that are not particularly useful, where we get lost in the trees and forget we are in a forest). The grouped frequency distribution is what we need to see where we are in the forest so that we can get out safely. Recipe 7-2 teaches you how to create a grouped frequency distribution, and you learn that you have control over the interval width so that you can specify an effective class interval size for your distribution.

Solution

The range in ages of the Oscar-winning actors and actresses is 21–80. Let us establish 12 intervals from 20 to 80 by setting breaks at 5 years. First, establish breaks by using the seq() function. Then use the cut() function to separate the data into the intervals. Because we do not want any overlap, we close each interval on the left and leave it open on the right. This is accomplished by using the right = FALSE argument. You can use the trick of an extra pair of parentheses to create the table and print it at the same time.

```
> breaks <- seq(20, 80, by = 5)
> ageIntervals <- cut(oscars$years, breaks, right = FALSE)
> (ages <- table(ageIntervals))
ageIntervals
[20,25) [25,30) [30,35) [35,40) [40,45) [45,50) [50,55) [55,60) [60,65) [65,70)
      5      26      28      36      28      22      10       4      12       0
[70,75) [75,80)
      1       1
```

Recipe 7-3. Calculating Summary Statistics

Problem

Frequency distributions are helpful, but they still leave us with unanswered questions about our data and its properties. These questions include the precise values for measures of the central tendency, variability, skewness, and kurtosis of a dataset. In particular, if data consist of more than one variable, tables become less useful, and we need statistics such as chi-square or likelihood ratios to describe the relationships among the variables. In Recipe 7-3, you learn how to use R base, as well as the contributed package fBasics to find the these summary statistics. We also explore the usefulness of the prettyR package for describing numerical data, as well as for frequency tables and crosstabulations.

Solution

Base R provides most of the common statistical indexes, but a few are missing. For example, though I have in my possession a book that claims the mode() function locates the modal value in a dataset, that is clearly not the case. The mode() function, as you have already seen, shows the storage class of an R object.

Table 7-1 lists the commonly used statistical functions in the base R distribution, which does not have functions for the mode, skewness, or kurtosis.

Table 7-1. *Commonly Used Statistical Functions in R*

Function	Calculates This
mean()	Arithmetic average
median()	Median
min()	Minimum value
max()	Maximum value
range()	Shows minimum and maximum values
IQR()	Interquartile range
sd()	Standard deviation, treating data as a sample
var()	Variance, treating data as a sample
length()	Counts the number of data points
sum()	The total of the data points
summary()	Five-number summary plus the mean

Here are the summary statistics for the 174 ages in the oscars data frame.

```
> mean(oscars$years)
[1] 40.08046
> median(oscars$years)
[1] 38
> min(oscars$years)
[1] 21
> max(oscars$years)
[1] 80
> range(oscars$years)
[1] 21 80
> IQR(oscars$years)
[1] 13.75
> sd(oscars$years)
[1] 11.05396
> var(oscars$years)
[1] 122.19
> length(oscars$years)
[1] 174
> sum(oscars$years)
[1] 6974
> summary(oscars$years)
   Min. 1st Qu.  Median    Mean 3rd Qu.    Max.
  21.00   32.00   38.00   40.08   45.75   80.00
```

A histogram shows that the ages are positively skewed (see Figure 7-1). Use the breaks argument to specify five-year intervals for the histogram, as follows:

```
> hist(oscars$years, breaks = seq(20,80, by = 5))
```

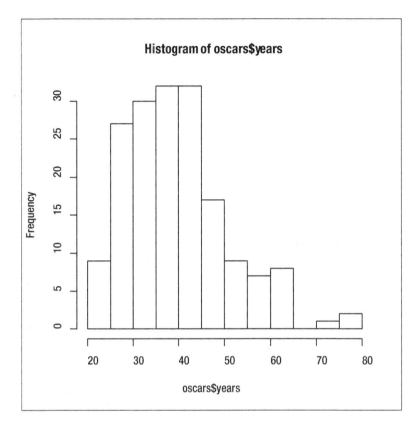

Figure 7-1. *The age distribution is positively skewed*

Many R packages add statistical functionality missing from the base version of R. One of the best of these is the fBasics package. Install fBasics and use the package for summary statistics.

```
> install.packages("fbasics")
> library(fBasics)
Loading required package: MASS
Loading required package: timeDate
Loading required package: timeSeries

Attaching package: 'fBasics'

The following object is masked from 'package:base':

    norm

> basicStats(oscars$years)
          X..oscars.years
nobs            174.000000
NAs               0.000000
Minimum          21.000000
Maximum          80.000000
```

```
1. Quartile          32.000000
3. Quartile          45.750000
Mean                 40.080460
Median               38.000000
Sum                6974.000000
SE Mean               0.837999
LCL Mean             38.426442
UCL Mean             41.734477
Variance            122.190021
Stdev                11.053959
Skewness              0.876438
Kurtosis              0.794613
```

This summary confirms the positive skew in the distribution of ages. The skewness coefficient would be zero if the data were completely symmetrical. Similarly, the data are positively *kurtotic*, meaning the distribution of ages is "peaked." In addition to the coefficients for skewness and kurtosis, we get the number of missing values (in this case none), the standard summary descriptive statistics, the standard error of the mean, and the lower and upper limits of a 95% confidence interval for the mean. In Chapter 12, you will learn some contemporary statistical methods for dealing with data such as these that depart from a normal distribution.

Another useful package is prettyR. It provides many options, including formatting R output for web display, the missing mode function we discussed earlier, and the ability to describe various data objects, including numeric and logical values.

```
> library(prettyR)
> describe(oscars$years)
Description of structure(list(x = c(22, 37, 28, 63, 32, 26, 31, 27, 27, 28, 30, 26, 29, 24, 38, 25,
29, 41, 30, 35, 32, 33, 29, 38, 54, 24, 25, 48, 41, 28, 41, 39, 29, 27, 31, 38, 29, 25, 35, 60, 61,
26, 35, 34, 34, 27, 37, 42, 41, 36, 32, 41, 33, 31, 74, 33, 49, 38, 61, 21, 41, 26, 80, 42, 29, 33,
36, 45, 49, 39, 34, 26, 25, 33, 35, 35, 28, 30, 29, 61, 32, 33, 45, 29, 62, 22, 44, 44, 41, 62, 53,
47, 35, 34, 34, 49, 41, 37, 38, 34, 32, 40, 43, 48, 41, 39, 49, 57, 41, 38, 39, 52, 51, 35, 30, 39,
36, 43, 49, 36, 47, 31, 47, 37, 57, 42, 45, 42, 45, 62, 43, 42, 48, 49, 56, 38, 60, 30, 40, 42, 37,
76, 39, 53, 45, 36, 62, 43, 51, 32, 42, 54, 52, 37, 38, 32, 45, 60, 46, 40, 36, 47, 29, 43, 37, 38,
45, 50, 48, 60, 50, 39, 55, 44)), .Names = "x", row.names = c(NA, -174L), class = "data.frame")

Numeric
             mean    median      var      sd   valid.n
x           40.08  .     38    122.2   11.05       174

> Mode(oscars$years)
[1] "41"
> mode(oscars$years)
[1] "numeric"
```

Remember R's case sensitivity. The Mode() function in prettyR finds the mode, while the mode() function returns the storage class. The prettyR package also "prettifies" frequency tables and crosstabulations. It also provides a very useful function to break down a numeric variable by one or more grouping variables. Let us examine the brkdn() function first. You must enter a formula showing the variable to be analyzed, the grouping variable, the data source, the number of grouping levels, the numeric description you want, and the number of places to round the results.

The formatting is very effective. Examine the means and variances for the ages of the male and female Oscar winners. It appears there may be some age discrimination against females.

```
> require(prettyR)
> brkdn(years ~ award, data = oscars, maxlevels = 2, num.desc = c("mean", "var"),
+ vname.width = NULL, width = 10, round.n = 2)

Breakdown of years by award
Level       mean        var
Actor       44.03       78.08
Actress     36.13       136.1
```

An independent-samples *t* test confirms the suspicion that the average ages for male and female winners are significantly different.

```
> t.test(oscars$years ~ oscars$award)
Welch Two Sample t-test

data: oscars$years by oscars$award
t = 5.0402, df = 160.244, p-value = 1.244e-06
alternative hypothesis: true difference in means is not equal to 0
95 percent confidence interval:
 4.809494 11.006598
sample estimates:
  mean in group Actor mean in group Actress
             44.03448              36.12644
```

Now, look at a frequency distribution and a crosstabulation formatted by prettyR. Because the age range is so great for the entertainers' ages, we will us a smaller dataset. See how much more informative the table produced by freq() in prettyR is than the one from the table() function is?

```
> freq(studentComplete$Quiz1)
```

```
Frequencies for studentComplete$Quiz1
        89  92  75  76  83  79  82  84  85  86  87  88  90  NA
         3   3   2   2   2   1   1   1   1   1   1   1   1   0
%       15  15  10  10  10   5   5   5   5   5   5   5   5   0
%!NA    15  15  10  10  10   5   5   5   5   5   5   5   5
```

```
> table(studentComplete$Quiz1)
```

```
75 76 79 82 83 84 85 86 87 88 89 90 92
 2  2  1  1  2  1  1  1  1  1  3  1  3
```

Using the plagiarism data frame from Chapter 6 (see Recipe 6-1), we crosstabulate the student's sex and whether he or she plagiarized the first assignment in the class. Compare the output from the prettyR calculate.xtab() function to that of the table() function in base R. It is obvious that prettyR wins that comparison.

> **calculate.xtab(plagiarism$Sex,plagiarism$Plagiarized1)**
Crosstabulation of plagiarism$Sex by plagiarism$Plagiarized1
        plagiarism$Plagiarized1

...	No	Yes	
F	27	4	31
	87.1	12.9	-
	67.5	80	68.89
M	13	1	14
	92.86	7.14	-
	32.5	20	31.11
	40	5	45
	88.89	11.11	100

odds ratio = 0.52
relative risk (plagiarism$Sex-M) = 0.55

> **table(plagiarism$Sex,plagiarism$Plagiarized1)**

```
    No Yes
 F 27   4
 M 13   1
```

Although the calculate.xtab() function is more informative than the table() function, the output is still not what you might see from the pivot table command in a commercial spreadsheet like Excel. The delim.xtab() function formats a crosstabulation with counts and percentages. You can specify whether to display the column of percentages next to the row of counts, or to display them separately. We will format the crosstabulation we just produced. By setting the interdigitate argument to F or FALSE, you will get the rows of counts and totals separately from the rows and totals of percentages—a more appealing alternative to me, at least, in terms of the visual appearance of the tables.

> **delim.xtab(crosstab, interdigitate = F)**
crosstab

	No	Yes	Total
F	27	4	31
M	13	1	14
Total	40	5	45

crosstab

	No	Yes	Total
F	87.1%	12.9%	100%
M	92.9%	7.1%	100%
Total	88.9%	11.1%	100%

For purposes of comparison, I show the pivot table produced by Excel 2013 for the same data, followed by the output from SPSS (see Figures 7-2 and 7-3).

Count of Sex	Plagiarized1		
Sex	No	Yes	Grand Total
F	27	4	31
M	13	1	14
Grand Total	40	5	45

Figure 7-2. *A pivot table in Excel 2013*

Sex * Plagiarized1 Crosstabulation

Count

		Plagiarized1		
		No	Yes	Total
Sex	F	27	4	31
	M	13	1	14
Total		40	5	45

Figure 7-3. *SPSS output of the crosstablulation*

Recipe 7-4. Working with Quantiles

Problem

We often find it useful to use intervals to represent the separation of data into more or less distinct categories. The most easily recognized of these is the percentile. By dividing a distribution into smaller groups of equal sizes, we are better able to visualize the location of an individual score in the data distribution. *Quantile* is a rather strange word, and not one on the tips of most people's tongues. Technically, quantiles are points taken at regular intervals from the cumulative density function (CDF) of a random variable. If we have *q* quantiles, we will thus divide the dataset into *q* equally sized subsets. Percentiles (also known as *centiles*) cut a distribution into 100 equal groups. Another common example is *quartiles*, which as the name implies, cut a distribution into fourths. The usefulness of these breakdowns is that they give us an idea of "how high" or "low" a given score is. The first quartile is also the 25th percentile, the second quartile is the 50th percentile (or the median), and the third quartile is the 75th percentile.

Solution

Although the definition of a quantile is fairly simple, the calculation of one is a bit problematic. For example, there are five different methods of determining quartiles in SAS. R's own `quantile()` function has nine different definitions (types) for calculating quantiles. Type `?quantile` at the command prompt to learn more about these types. The default is type 7, which is consistent with that of the S language. When I was an R newbie, I was perplexed that my TI-84 calculator, Excel, SPSS, Minitab, and R were inconsistent in reporting values for quartiles. Since then, I have learned a lot more about R, and Microsoft has even patched up its quartile function so Excel produces the correct values according to the National Institute of Standards and Technology (NIST).

Continuing with our data on Oscar winners, let's look at the quartiles. Observe the use of a vector of probabilities between 0 and 1 to establish which quantiles to report. We could just as easily report deciles (tenths) if we desired to. R reports the minimum value as the 0th percentile and the maximum value as the 100th percentile. Even that is a matter of controversy, as many statisticians (myself among them) argue that the 0th and 100th percentiles are undefined.

```
quantile(oscars$years, prob = seq(0, 1, 0.25))
    0%    25%    50%    75%   100%
21.00  32.00  38.00  45.75  80.00
> quantile(oscars$years, prob = seq(0, 1, 0.10))
   0%   10%   20%   30%   40%   50%   60%   70%   80%   90%  100%
21.0  27.3  30.0  33.0  36.0  38.0  41.0  44.0  48.4  55.7  80.0
```

CHAPTER 8

■ ■ ■

Graphics and Data Visualization

In Chapter 8, you learn how to graph and visualize data. As in other areas, R's base graphics capabilities are fine for most day-to-day graphing purposes, but a number of packages extend R's graphical functionality. We will use the ggplot2 package to illustrate.

The ggplot2 package was written by Hadley Wickham and Winston Chang as an implementation of the "grammar of graphics." This term was introduced by Leland Wilkinson in his book on the subject published by Springer in 2005. The idea was to take the good parts of the base R graphics and lattice graphics, and none of the bad parts. All the graphics you have seen thus far were created using the base version of R.

A point about graphics and visualization is in order. The two are not exactly the same thing. A graph is a kind of diagram that shows the relationships between two or more things represented as dots, bars, or lines. To *visualize* something, on the other hand, we first develop a mental picture. The image in our mind helps us understand the data in a way that looking at raw data cannot do. Of course, what has happened is that visualization has evolved to include not just a mental process, but external processes as well. We could say that visualization is correctly described as turning something that is symbolic into something geometric. Visualization allows researchers to "observe" their own simulations and calculations.

It is very easy to make bad graphs. It is not very easy to make beautiful ones. If you would like to sit at the feet of a master, study the work of Edward Tufte. His books on the visual display of quantitative data are masterpieces. According to Tufte, "Excellence in statistical graphics consists of complex ideas communicated with clarity, precision, and efficiency," (from the second edition of *The Visual Display of Quantitative Information*, Graphics Press, 2001). Tufte's book has approximately 250 illustrations of very good graphics and a few examples of really terrible ones. My favorite thing about Tufte is that he dislikes PowerPoint as much as I do (www.edwardtufte.com/tufte/powerpoint).

We begin with the ubiquitous pie chart. According to Wilkinson, the pie chart has been praised unjustifiably by managers and maligned unjustifiably by statisticians. For example, the R documentation for the pie() function states: "Pie charts are a very bad way of displaying information. The eye is good at judging linear measures and bad at judging relative areas. A bar chart or dot chart is a preferable way of displaying this type of data." I agree with the second and third sentences, but not with the first one. In addition to the pie chart, we will run through the standard graphics, some of which you have already seen in earlier chapters. We will also explore several techniques for visualizing data, including categorical and scale data.

Recipe 8-1. Getting the Colors You Want
Problem

R's default color palette is a series of pastel colors, much like the sidewalk chalk children like to play with. The colors are not very appealing, but they are functional for many basic uses. Users often want to establish a more effective color scheme. You learn several important skills in Recipe 8-1, including choices for color palettes, how to create a pie chart representation of a color palette, and how to display multiple graphic objects simultaneously.

Solution

The pie chart is so simple that a five-year-old can grasp the concept. It is also so appealing to the eye that people cannot resist making pie charts. It is also very easy to muck one up, and people do it all the time. One thing you should never do is to make a 3D pie chart. Excel will readily do this for you, but a 3D pie chart is a definite no-no. The 3D pie chart distorts the perspective of the data as well as introduces a false third dimension. R makes decent pie charts, and they are completely customizable. My guess is that Hadley Wickham doesn't like pie charts. The ggplot2 package features many "geoms," but there is not one for a pie chart.

■ **Note**   The label "geom" is short for *geometric object*.

One excellent use for a pie chart is to explore the color palettes in R. I wish I could say I thought this up, but the example is in the R documentation for the palette function. We first draw a blank pie chart by using the argument col = FALSE, and then fill the graphics device with the blank pie chart, R's default palette, the rainbow palette (my personal favorite), and the heat colors palette. Here's the code for doing that. The par() function allows the user to place multiple graphic objects in the same window, in this case, four objects arranged in two rows and two columns, but with no borders displayed. The argument main can be used to specify the title of each separate graphic object.

```
    n<- 7
par(mfrow = c(2,2))
pie(rep(1,n), col = FALSE, main = "Blank Pie")
pie(rep(1,n), main = "Default Colors")
pie(rep(1,n), col = rainbow(n), main = "Rainbow Colors")
pie(rep(1,n), col = heat.colors(n), main = "Heat Colors")
```

See Figure 8-1 for the graphical representations.

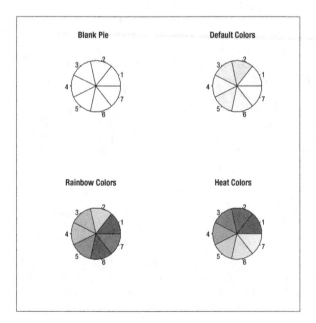

Figure 8-1. *Illustration of a few of R's color palettes*

In addition to learning about the color palettes, you just learned how to display multiple graphs in the R Graphics Device. Examine Figure 8-2 to see that you can copy and save the output, as well as print it. To access these options, just right-click anywhere in the Graphics Device window.

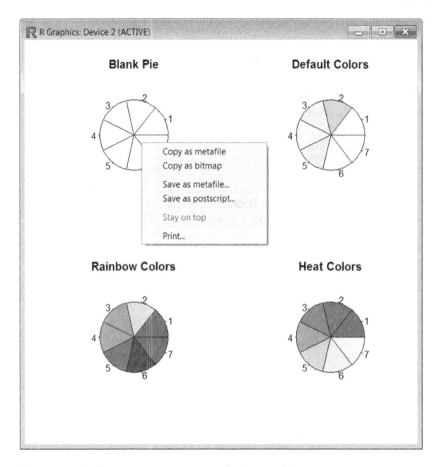

Figure 8-2. *R allows you to copy, save, and print graphics*

Recipe 8-2. Using the Standard Graphs
Problem

Many people are visual learners and thinkers. Numbers are wonderful, and I like playing with numbers and "crunching" them, but I really like to turn them into pictures, both mental and physical. The trite expression that a picture is worth a thousand words has a ring of truth to it.

Every statistics book in my library has a chapter on the graphical display of data. This is usually the second or third chapter in the book, because graphs are part of descriptive statistics, or what Tukey called *exploratory data analysis*. You have seen frequency distributions, histograms, pie charts, and bar charts (including clustered bar charts) in earlier chapters. In the remaining recipes in Chapter 8, you learn to produce line graphs, scatterplots, and various exploratory graphs and plots. We will use ggplot2 for most of the graphics in this chapter.

Solution

First, let us review some principles of good graphics before illustrating the various graphs that I just listed. Here are Tufte's principles of graphical integrity (www.asq0511.org/Presentations/1298/sld008.htm):

- The physical area on the graphic should be directly proportional to the number represented.

- The data and important events should be labeled and explained on the graphic.

- The graphic should show variation in the data, not design variation.

- Time series displays of money should use deflated and standardized units.

- The number of chart dimensions should not exceed the number of data dimension.

- Do not quote data out of context.

Histograms

The bars in a histogram represent the frequencies of the observations. Because there is an underlying continuum, the bars should touch. Histograms are one of the most common and most meaningful ways to display simple and grouped frequency distributions.

Here is an example of a histogram taken from the ggplot2 documentation. In keeping with Tufte's principles, I added a descriptive title. The code follows. The data for more 58,788 movies ship with ggplot2. The rating is based on International Movie Database (IMDB) user votes and can range from 1 to a perfect 10. We take a sample of 1,000 to keep our example manageable. The set.seed function allows us to generate pseudorandom numbers that we can replicate should we want to use the same example again. This is useful for various purposes; in this case, for sampling from a larger dataset. See Figure 8-3 for the histogram.

```
> library(ggplot2)
> set.seed(5689)
> movies <- movies[sample(nrow(movies), 1000), ]
> qplot(rating, data=movies, geom="histogram", main = "Movie Ratings")
```

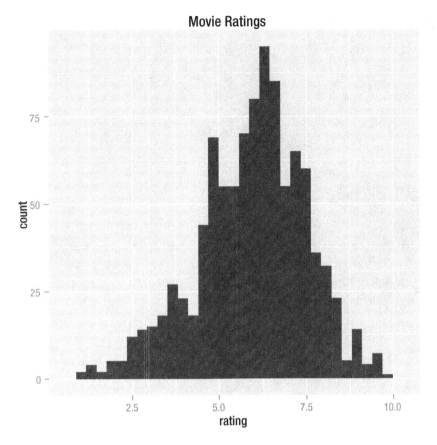

Figure 8-3. *Histogram produced by ggplot2*

Bar Charts

Both histograms and bar charts are produced by the geom_bar() function in ggplot2. Here are some actual expense data from one of my recent conference trips. First I create a data frame called df, which contains the meal and prices. Then I download ggplot2 using the command install.packages("ggplot2"), followed by library("ggplot2") after the download is completed.

The aes() function creates the "aesthetic" for the bar chart. The default legend is redundant in this case, but can be removed by setting guides(fill = FALSE). We define the bar chart by telling R what the axes are, as well as what variable to fill. The stat = "identity" option specifies that that the bars should represent the actual values in the data frame. The default is stat = "bin", which causes the bars to represent frequencies. The completed bar chart appears in Figure 8-4.

```
> df
        Meal Price
1 Breakfast 13.98
2      Lunch 21.52
3     Dinner 37.52
> barChart<- ggplot(df)
> barChart <- barChart +geom_bar(aes(x=Meal,y=Price, fill=Meal), stat ="identity")
> barChart + ggtitle("Meal Expenses")
> barChart + guides(fill = FALSE)
```

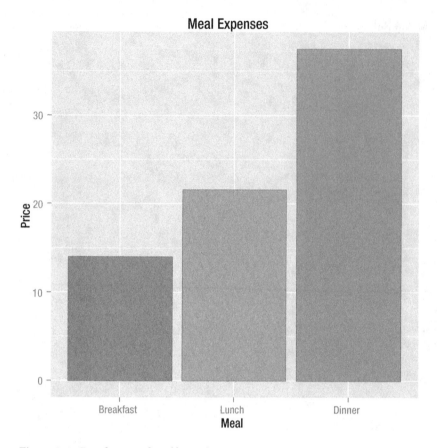

Figure 8-4. *Bar chart produced by ggplot2*

Line Graphs

We use line graphs to display the change in a variable or variables over time. A common example is the change in the closing price of a stock. A frequency polygon is also a kind of line graph. The following data are from the Yahoo! Finance web site. The data include the opening, high, low, closing, volume, and adjusted closing prices of Starbucks stock for the year 2013. I added an index number for the 252 days in the dataset.

```
> head(starbucks)
      date  open  high   low close  volume adj_close number
1 1/2/2013 54.59 55.00 54.26 55.00 6633800    53.88      1
2 1/3/2013 55.07 55.61 55.00 55.37 7335200    54.24      2
3 1/4/2013 55.53 56.00 55.31 55.69 5455700    54.55      3
4 1/7/2013 55.40 55.79 55.01 55.72 4360000    54.58      4
5 1/8/2013 55.58 55.72 55.07 55.62 4806700    54.48      5
6 1/9/2013 55.89 55.90 54.33 54.63 8339200    53.51      6
```

We can use the qplot() function in ggplot2 for most basic graphs. See the syntax as follows. Note that the geom type is specified as a character string. The finished line graph appears in Figure 8-5.

```
> qplot(starbucks$number, starbucks$adj_close, geom="line", main="Starbucks Closing Price")
```

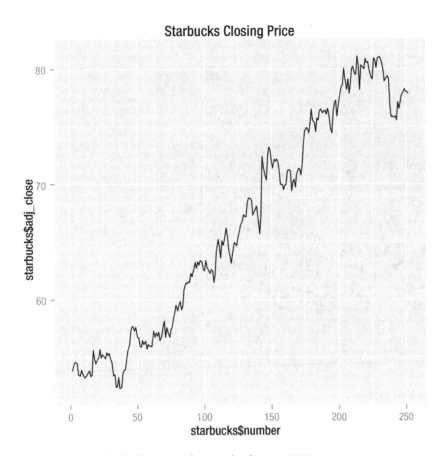

Figure 8-5. *Starbucks closing stock prices for the year 2013*

Scatterplots

The scatterplot is the default in ggplot2 if no geom is specified. This is consistent with the base version of R, which will produce a scatterplot when you plot two variables without further specification. Let us see how the Starbucks volumes relate to the date. Remember the xlab and ylab arguments are used to specify the labels for the *x* and *y* axes.

```
> qplot(starbucks$number, starbucks$volume, xlab = "date", ylab = "volume", main = "Starbucks Volume by Date")
```

The scatterplot appears in Figure 8-6. There is no apparent relationship between the date and the volume.

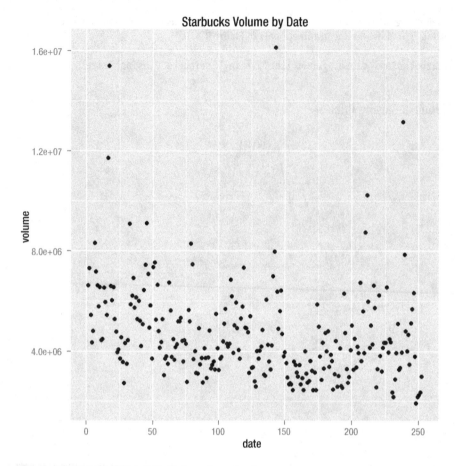

Figure 8-6. *Scatterplot of Starbucks volume by date*

Recipe 8-3. Using Graphics for Exploratory Data Analysis

Problem

According to John Tukey's preface to *Exploratory Data Analysis* (Pearson, 1977): "Once upon a time, statisticians only explored. Then they learned to confirm exactly—to confirm a few things exactly, each under very specific circumstances. As they emphasized exact confirmation, their techniques inevitably became less flexible."

Tukey believed that exploratory data analysis is "detective work," and I agree. The two most common graphical tools for exploratory data analysis are the stem-and-leaf plot and the boxplot. These tools do help you see another, deeper layer in the data. The boxplot, in particular, is very useful for visualizing the shape of the data, as well as the presence of outliers. In addition to these plots, we will examine the dot plot (of which there are a couple of varieties).

Solution

A boxplot is a visual representation of the Tukey five-number summary of a dataset. These numbers are the minimum, the first quartile, the median, the third quartile, and the maximum. Tukey originally called the plot a "box and whiskers" plot.

Side-by-side boxplots are very handy for comparing two or more groups. The following example uses the PlantGrowth data frame that comes with the base version of R. The data represent the weights of 30 plants divided into three groups on the basis of a control condition and two treatment conditions:

```
> summary(PlantGrowth)
     weight        group
 Min.   :3.590   ctrl:10
 1st Qu.:4.550   trt1:10
 Median :5.155   trt2:10
 Mean   :5.073
 3rd Qu.:5.530
 Max.   :6.310
```

Here is how to make side-by-side boxplots for the three groups using qplot(). See the graphics output in Figure 8-7, where you will note the presence of two outliers in the trt1 group.

```
> boxPlot <- ggplot(PlantGrowth, aes(x=group, y=weight)) + geom_boxplot()
> boxPlot
```

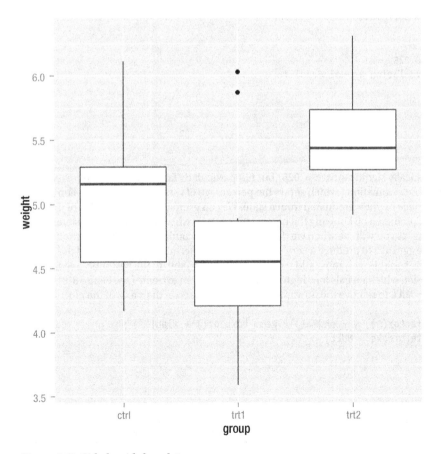

Figure 8-7. *Side-by-side boxplots*

I retrieved the following information from the Internet. The data include the location (the 50 US states plus Washington, DC), the region, the 2013 population, the percentage of the population with PhDs, the per capita income in 2012, the total number of PhDs awarded in that state or location during 2012, and the breakdown of these degrees by male and female. The data are part of a much larger database managed by the National Science Foundation.

```
> head(phds)
     location         region  pop2013 phdPct income total maleTot femaleTot
1     Alabama      southEast  4833722 0.0134  35625   648     328       320
2      Alaska nonContiguous   735132 0.0068  46778    50      29        21
3     Arizona           west  6626624 0.0134  35979   888     491       397
4    Arkansas      southEast  2959373 0.0066  34723   194     110        84
5  California           west 38332521 0.0157  44980  6035    3423      2612
6    Colorado        central  5268367 0.0154  45135   809     427       382
```

It would be interesting to see if there is a positive correlation between the percentage of the population with PhDs and the per capita income. There is.

```
> cor.test(phds$phdPct, phds$income)

        Pearson's product-moment correlation

data: phds$phdPct and phds$income
t = 2.3124, df = 49, p-value = 0.025
alternative hypothesis: true correlation is not equal to 0
95 percent confidence interval:
 0.0416852 0.5423665
sample estimates:
      cor
0.3136656
```

The correlation of $r = .31$ is statistically significant at $p = .025$, but not particularly large. Squaring the correlation coefficient gives us the "coefficient of determination," which shows the percentage of variance overlap between the per capita income and the percentage of PhDs graduated from a state. We can estimate that about 9.8% of the variation in per capita income can be predicted by knowing the percentage of PhDs who graduated in that state. However, the correlation is misleading, as you will see when we determine there are outliers throwing off the relationship. As indicated, we can use ggplot2 to produce a very nice box plot to determine the presence of the outliers. To make a boxplot for a single variable, you must add a fake x grouping variable to the aesthetic. See the following code and the resulting boxplot, which reveals four high outliers, one of them extreme (see Figure 8-8). We use scale_x_discrete(breaks = NULL to remove the unwanted zero that appears on the x axis of the plot.

```
> plot <- ggplot(phds, aes(x = factor(0), y = phdPct))+ geom_boxplot() + xlab("")
> plot <- plot + scale_x_discrete(breaks = NULL)
> plot
```

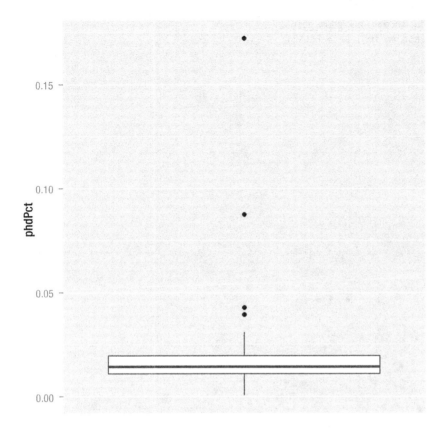

Figure 8-8. *Boxplot reveals four outliers*

The four locations with the highest percentages of PhDs are the District of Columbia, Massachusetts, Washington, and Wisconsin. With these locations taken out of the data, the correlation becomes slightly higher:

```
> cor.test(data = phds, x = phds$phdPct, y = phds$income)

        Pearson's product-moment correlation

data: phds$phdPct and phds$income
t = 2.4334, df = 44, p-value = 0.01909
alternative hypothesis: true correlation is not equal to 0
95 percent confidence interval:
 0.06011174 0.57700935
sample estimates:
      cor
0.3443999
```

A scatterplot also reveals that with the outliers removed, there is basically a linear relationship between the percent of PhDs and the per capita income in a state. We can still account for only 11.9% of the variation in income, however (see Figure 8-9, in which I used ggplot to add the line of best fit). The inclusion of geom_point() adds the required layer to the scatterplot to make it visible in the R Graphics Device. The fit line is added by stat_smooth, which plots a smoother—in this case, a linear model— on the scatterplot.

```
> plot <- ggplot(phds,aes(phdPct, income)) + geom_point()
> plot <- plot + stat_smooth(method = "lm", se = FALSE)
> plot
```

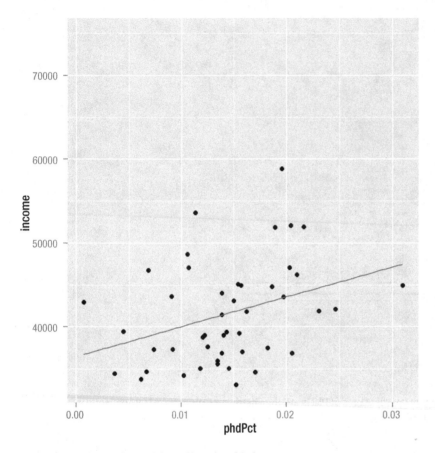

Figure 8-9. *Scatterplot with line of best fit added*

Stem-and-leaf plots are a semigraphical way to show the frequencies in a dataset. The stems are the leading digit(s) and the leaves are the trailing digits. A stem-and-leaf plot (also known as a *stemplot*) shows every value in the dataset, preserving what I call its "granularity." The stem-and-leaf plot is not available in ggplot2, but it is in the base version of R.

One of my favorite R datasets is faithful, which lists the waiting time between eruptions and the duration of the eruptions of the Old Faithful geyser in Yellowstone National Park. Let us produce a stem-and-leaf plot of the waiting times. Because there are so many observations, R created separate bins for the trailing digits 0–4 and for 5–9 in a split stem-and-leaf plot. Clearly, if the stem-and-leaf plot were rotated 90 degrees counterclockwise, it would resemble a grouped frequency histogram. Unlike other graphs, the stem-and-leaf plot is displayed in the R Console.

```
> stem(faithful$waiting)

  The decimal point is 1 digit(s) to the right of the |
```

```
4 | 3
4 | 55566666777788899999
5 | 000001111112222233333334444444444
5 | 5555556666677788889999999
6 | 00000022223334444
6 | 555667899
7 | 0000111123333333444444
7 | 555555555666666666777777777777788888888888888889999999999
8 | 000000000111111111111112222222222222233333333333333334444444444
8 | 555555666666778888888999
9 | 00000012334
9 | 6
```

Dotplots are another very good way to preserve the granularity of the dataset. The newest release of the ggplot2 package has dotplots, but, at least according to Twitter, will not have stem-and-leaf plots any time soon. As it materializes, there are different kinds of dotplots. The frequency dotplot was described by Leland Wilkinson, the mastermind behind the grammar of graphics. There is also a version of the dotplot developed by William Cleveland as an alternative to bar charts. See the completed Wilkinson dotplot in Figure 8-10. For each aesthetic used to create a plot in ggplot, one must supply the proper "geom" to produce the required layer.

```
> ggplot(faithful, aes(x = waiting)) + geom_dotplot()
stat_bindot: binwidth defaulted to range/30. Use 'binwidth = x' to adjust this.
```

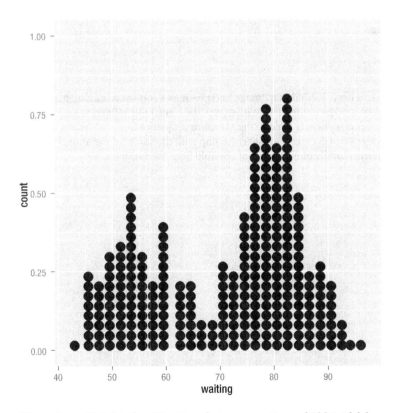

Figure 8-10. *Dotplot of waiting times between eruptions of Old Faithful*

Recipe 8-4. Using Graphics for Data Visualization

Problem

As was mentioned in the introduction to this chapter, data visualization is not the same thing as a graph. Certain graphs discussed in the other recipes in this chapter are useful for visualization. In Recipe 8-4, for example, boxplots and scatterplots can quickly tell us something about the relationship between variables and about the presence and the effect of outliers. Similarly, there are many other diagnostic plots that help us delve deeper into the unseen properties of the data.

Data visualization is becoming increasingly important, and also includes the visualization of ideas and concepts. For example the word cloud is an interesting way to visualize the occurrences of words or references in a text or a web site. We will cover word clouds in Chapter 14, but for now, let's use ggplot2 and learn how to produce maps, and then populate them with data to help us visualize the information more effectively.

Solution

R is an effective visualization tool with the ggplot2 package installed. One of the best uses of ggplot2 is to create maps that help to visualize one or more attributes of a geographical location. According to Hadley Wickham, the grammar of graphics requires that every plot must consist of five components:

- a default dataset with aesthetic mappings

- one or more layers, each with a geometric object, a statistical transformation, and a dataset with aesthetic mappings

- a scale for each aesthetic mapping

- a coordinate system

- a facet specification

The purpose of the ggmap package is to take a downloaded map image and plot it as a context layer in ggplot2. The user can then add layers of data, statistics, or models to the map. With ggmap, the *x* aesthetic is fixed to longitude, the *y* aesthetic is fixed to latitude, and the coordinate system is fixed to the Mercator projection.

Compare a table of the level of unemployment by US county in 2009 to a map with the same information (see Figure 8-11). It is obvious that the map allows visualization that is not possible from the raw data.

```
> head(unemp)
   fips   pop unemp colorBuckets
1 1001 23288   9.7            5
2 1003 81706   9.1            5
3 1005  9703  13.4            6
4 1007  8475  12.1            6
5 1009 25306   9.9            5
6 1011  3527  16.4            6
> tail(unemp)
        fips   pop unemp colorBuckets
3213 72143 12060  16.3            6
3214 72145 20122  17.6            6
3215 72147  3053  27.7            6
3216 72149  9141  19.8            6
3217 72151 11066  24.1            6
3218 72153 16275  16.0            6
```

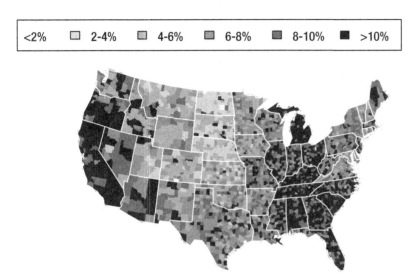

unemployment by county, 2009

| <2% | □ 2-4% | ▢ 4-6% | ▨ 6-8% | ▩ 8-10% | ■ >10% |

Figure 8-11. *Unemployment by county, 2009*

The following example is based on the Harvard University Center for Computational Biology's course on spatial maps and geocoding in R (`http://bcb.dfci.harvard.edu/~aedin/courses/R/CDC/maps.html`). Here is how the map was constructed. Note the use of color to add visual appeal to the map. We cut the unemployment statistics into five categories, plot the map, and then "fill the bucket" for each US county. The result is a visually attractive and informative overview of unemployment at the state and county levels. FIPS are Federal Information Processing Standards five-digit codes assigned by the National Institute of Standards and Technology (NIST) to specify US states (the first two digits) and counties (the last three digits).

```
data(unemp)
data(county.fips)

# Plot unemployment by country
colors = c("#F1EEF6", "#D4B9DA", "#C994C7", "#DF65B0", "#DD1C77",
    "#980043")
head(unemp)
unemp$colorBuckets <- as.numeric(cut(unemp$unemp, c(0, 2, 4, 6, 8,
    10, 100)))
colorsmatched <- unemp$colorBuckets[match(county.fips$fips, unemp$fips)]

map("county", col = colors[colorsmatched], fill = TRUE, resolution = 0,
    lty = 0, projection = "polyconic")
## Loading required package: mapproj
```

```
# Add border around each State
map("state", col = "white", fill = FALSE, add = TRUE, lty = 1, lwd = 0.2,
    projection = "polyconic")
title("unemployment by county, 2009")

leg.txt <- c("<2%", "2-4%", "4-6%", "6-8%", "8-10%", ">10%")
legend("topright", leg.txt, horiz = TRUE, fill = colors)
```

Of course, we can visualize other things besides unemployment statistic by US county. For example, here is a map of the southeastern United States (see Figure 8-12). I used ggplot to build the map, and the geocode() function to look up the latitude and longitude values for a state university in each state. I then looked up the fall 2013 enrollment for each school and plotted the schools' locations along with a label and the total enrollment, which shows as a larger filled circle for larger enrollments. Here is the code that accomplished this. Only a couple of the geocodes are shown to illustrate.

```
> geocode("university of georgia")
Information from URL : http://maps.googleapis.com/maps/api/geocode/json?address=
university+of+georgia&sensor=false
Google Maps API Terms of Service : http://developers.google.com/maps/terms
        lon      lat
1 -83.37732 33.94801
> geocode("university of alabama")
    Information from URL : http://maps.googleapis.com/maps/api/geocode/json?address=
university+of+alabama&sensor=false
Google Maps API Terms of Service : http://developers.google.com/maps/terms
        lon      lat
1 -87.54743 33.21444

> mydata
                                       school      long      lat enrollment    label
1                       University of Alabama -87.54743 33.21444      34852       UA
2                        University of Florida -82.34639 29.64526      49042       UF
3                        University of Georgia -83.37732 33.94801      34536      UGA
4                       University of Kentucky -84.50397 38.03065      28037       UK
5                    University of Mississippi -89.53844 34.36473      22286 Ole Miss
6 University of North Carolina at Chapel Hill -79.04691 35.90491      28136      UNC
7                University of South Carolina -81.02743 33.99611      32848      USC
8       University of Tennessee at Knoxville -83.92074 35.96064      27171      UTK
9                       University of Virginia -78.50798 38.03355      23464      UVA
```

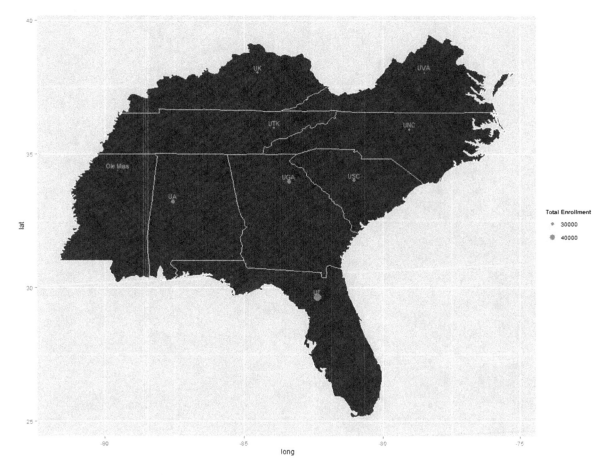

Figure 8-12. *Universities in the southeast and their enrollments*

Now, here is the code for making the map:

```
library(ggplot2)
library(maps)
#load us map data
all_states <- map_data("state")
southeast <- subset(all_states, region %in% c("florida", "georgia", "south carolina",
"north carolina", "virginia","kentucky","tennessee","mississippi","alabama"))
p <- ggplot()
p <- p + geom_polygon(data=southeast, aes(x=long, y=lat, group = group), color="white" )
p <- p + geom_point( data=mydata, aes(x=long, y=lat, size = enrollment), color="coral1") +
scale_size(name="Total Enrollment")
p <- p + geom_text( data=mydata, hjust=0.5, vjust=-0.5, aes(x=long, y=lat, label=label),
color="gold2", size=4 )
```

And finally, the map itself (see Figure 8-12).

CHAPTER 9

■ ■ ■

Probability Distributions

Looking up probabilities once was a time-consuming, complex, and error-prone task. Virtually every statistics book has tables of various probability distributions in an appendix, but R provides a family of functions for both discrete and continuous probability functions, making the use of tables unnecessary. You have already seen the rnorm() function used to produce random numbers that simulate a normal distribution with a certain mean and standard deviation. The functions for other probability distributions are all similarly labeled to make it easy to remember them.

In this chapter, you will learn how to use R for finding various probabilities for both discrete and continuous probability distributions. The normal distribution can serve as a model for the rest. The density function is dnorm(). The cumulative probability distribution (PDF) is pnorm() and the quantile function is qnorm(). The function for simulating a random value based on the normal distribution is rnorm().

Recipe 9-1. Finding Areas Under the Standard Normal Curve
Problem

Everyone who uses statistics on even a casual basis needs to look up the probability to the left or to the right of a given z score. It is also very common to need to find the area between two z scores. Because the standard normal distribution is symmetrical, many of the tables go from the mean of 0 to a high z score of 3.5 or 4.0. In Recipe 9-1, you learn how to build and plot your own standard normal distribution to visualize areas under the curve. You also learn how to "look up" areas and plot them for visualization.

Solution

First, we will build two vectors. One is the x axis (z scores ranging from –4 to 4) and the other is the probability density for each z score. This is the y axis. We will combine the vectors into a data frame, which we will abbreviate df. To get the densities, we use the dnorm() function.

```
> xaxis <- seq(-4, 4, by = 0.10)
> yaxis <- dnorm(xaxis)
> df <- as.data.frame(cbind(xaxis, yaxis))
```

We now have everything we need to build a standard normal distribution and plot it (see Figure 9-1). Here's how to build the curve. Though technically not required, combining the axes into a data frame allows us to use ggplot2 with the data frame to produce different graphics. Here, however, we simply use the base R graphics module to plot the normal curve. To clarify matters, the type is "l", which stands for "line," and not the numeral 1.

```
df <- as.data.frame(cbind(xaxis, yaxis))
> plot(df, xlab = "z score", ylab = "density", type = "l")
```

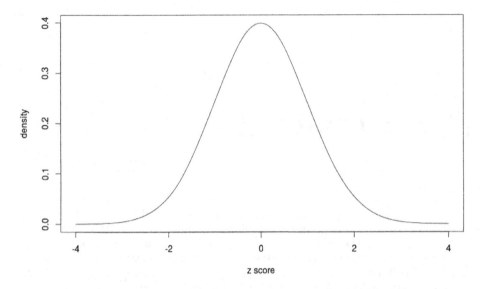

Figure 9-1. *The standard normal curve*

We see the familiar unimodal, symmetrical, bell-shaped curve and note that the probabilities taper off rapidly after values of $z = \pm 3$.

You can find areas under the normal curve by using a subtraction strategy. The pnorm() function also allows you to find both right- and left-tailed probabilities. You can find the area between any two z scores simply by subtracting the area to the left of the lower z score from the area to the left of the higher z score. Find the area to the left of a z score by using the pnorm() function. Similarly, you can find the area to the right of a given z score by setting the lower argument to FALSE (or by finding 1-pnorm()). Let us determine the areas to the left and right of a z score of 1.96:

```
> pnorm(1.96)
[1] 0.9750021
> pnorm(1.96, lower = FALSE)
[1] 0.0249979
```

Clearly, if we subtract the area to the left of $z = -1.96$ from the area to the left of $z = 1.96$, we will be left with the middle 95% of the standard normal distribution:

```
> pnorm(1.96) - pnorm(1.96, lower = FALSE)
[1] 0.9500042
We could also find this using pnorm(1.96)-pnorm(-1.96).
```

It is similarly easy to use the qnorm() function to locate critical values. Assume we want the value of z that cuts off the top 10% of the standard normal distribution from the lower 90%. We use the qnorm() function as follows:

```
> qnorm(.90)
[1] 1.281552
```

We can also supply a vector of quantiles to the qnorm() function. For example, here are the critical values of z for alpha levels of .10, .05, and .01.

```
> quantiles <- c(.95, .975, .995)
> qnorm(quantiles)
[1] 1.644854 1.959964 2.575829
```

Recipe 9-2. Working with Binomial Probabilities

The binomial probability distribution is the distribution of the random variable r, the number of "successes" in a series of N dichotomous (0,1) trials, where the result of each trial is a 0 (failure) or a 1 (success). Often the definition of success is arbitrary, as in heads or tails in a coin toss, but sometimes the definition of success is logical, as in a correct answer to a multiple-choice question or a unit without a defect. The binomial distribution gives the probability of r successes in N trials, where r can range from 0 to N. The most common and well-known example is that of a coin toss. We know intuitively that a fair coin has a probability of .5 of landing on heads, and a probability of .5 of landing on tails. Yet when we flip the coin, we get one or the other. The two outcomes are dbinom mutually exclusive. The law of large numbers tells us that over a long series of trials, the empirical probability of heads will converge on the theoretical probability.

Problem

Technically, the binomial probability distribution has a probability mass function (PMF), rather than a PDF, because the binomial distribution is discrete. Finding binomial probabilities is similar to finding normal probabilities.

Solution

The binomial distribution is used in a variety of applications. Binomial distributions have the following four characteristics:

- The probability of success, p, remains constant for all trials.

- The procedure has a fixed number of trials, N.

- The N trials are independent. The outcome of any trial does not affect the outcome of any other trial.

- Each trial must result in either a success or a failure.

As an example, assume that a manufacturing process is known to produce 4.7% defective parts. A random sample of 10 parts is obtained. What is the probability that exactly 3 parts are defective? What is the probability that 0 parts are defective? What is the probability that 4 or more parts are defective? The binomial distribution can help us answer these questions. For the sake of illustration, let us generate the entire distribution of binomial probabilities for exactly 0 to 10 defective parts given a defect rate of .047. We use the dbinom() function in this case. For convenience, we create a data frame to show the number of "successes" (in this case, defects) and the exact probabilities.

```
> r <- 0:10
> N <- 10
> p <- .047
> prob <- round(dbinom(r, N, p), 5)
> prob <- as.data.frame(cbind(r, prob))
> prob
```

```
    r    prob
1   0 0.61792
2   1 0.30474
3   2 0.06763
4   3 0.00889
5   4 0.00077
6   5 0.00005
7   6 0.00000
8   7 0.00000
9   8 0.00000
10  9 0.00000
11 10 0.00000
```

The probability of exactly 3 parts being defective is .06763. The probability of exactly 0 parts being defective is .61792. The probability of 4 or more parts being defective is .00077 + .00005 = .00083. We can also use the cumulative binomial distribution to solve the last problem, the probability of 4 or more parts being defective. The complement rule of probability says that we can find the complement of a probability p by subtracting p from 1. We can find the left-tailed probability of $r \mid R <= 3$ as follows, and then subtract that value from 1 to get the probability of 4 or more defects.

```
> round(1 - pbinom(3, 10, .047), 5)
[1] 0.00082
```

As another example, assume that a commercial airliner has four jet engines. The independent reliability of each engine is $p = .92$ or 92%. For the aircraft to land safely, at least two of the engines must be working properly. What is the probability that the flight can land safely? The probability that two or more engines will work properly is .998 or 99.8%. We can find this using the complement rule, or equivalently by setting the lower argument to F or FALSE:

```
> N <- 4
> p <- .92
> 1 - pbinom(1, N, p)
[1] 0.9980749
> pbinom(1, N, p, lower = F)
[1] 0.9980749
```

Recipe 9-3. Working with Poisson Probabilities
Problem

The Poisson distribution (named after French mathematician Siméon Denis Poisson) shows the number of "occurrences" within a given interval. The interval may be distance, time, area, volume, or some other measure. Unlike the binomial distribution, the Poisson distribution does not have a fixed number of observations. The Poisson distribution applies in the following circumstances:

- The number of occurrences that occur within a given interval is independent of the number of occurrences in any other nonoverlapping interval.

- The probability that an occurrence will fall within an interval is the same for all intervals of the same size, and is proportional to the size of the interval (for small intervals).

- As the size of the interval becomes smaller, the probability of an occurrence in that interval approaches zero.

Remember the Poisson counts have no theoretical upper bound, so we often use the complement rule of probability to solve Poisson problems. In many disciplines, certain events need to be counted or estimated in order to make good decisions or plans. For example, the number of cars going through an intersection, or the number of accidents at the intersection could be a justification for installing a traffic light. Estimating the number of calls arriving at a customer service center is important for staffing purposes.

Solution

The Poisson distribution is useful for determining probabilities of events that occur rarely, but are significant when they do occur. One of the earliest applications of the Poisson distribution was its use by Ladislaus Bortkiewicz in 1898 to study the rate of deaths among Prussian soldiers who had been accidentally kicked by horses.

In more routine applications, the Poisson distribution has been used to determine the probabilities of industrial accidents, hurricanes, and manufacturing defects. As an example, imagine that a manufacturing plant has historically averaged six "reportable" accidents per quarter. What is the probability of exactly six accidents in the next quarter? What is the probability of zero accidents? What is the probability of more than six accidents? We can use the functions for the Poisson distribution to answer these questions. We must supply the number of occurrences (which can be a vector) and lambda (which is the mean number of occurrences per interval). Following the pattern of binomial probabilities, dpois(x, lambda) gives the probability that there are x occurrences in an interval, while ppois(x, lambda) gives the probability of x or fewer occurrences in an interval, when the mean occurrence rate for intervals of that size is lambda.

```
> dpois(6, 6) ## The exact probability of 6 accidents
[1] 0.1606231
> dpois(0, 6) ## The exact probability of 0 accidents
[1] 0.002478752
> 1 - ppois (6, 6) ## The probability of more than 6 accidents
[1] 0.3936972
```

Recipe 9-4. Finding *p* Values and Critical Values of *t*, *F*, and Chi-Square
Problem

P values and critical values figure prominently in inferential statistics. Historically, tables of critical values were used to determine the significance of a statistical test and to choose the appropriate values for developing confidence intervals. Tables are helpful, but often incomplete, and the use of tables is often cumbersome and prone to error. You saw earlier that the qnorm() function can be used to find critical values of z for different alpha levels. We can similarly use qt(), qf(), and qchisq() for finding critical values of t, F, and chi-square for various degrees of freedom. The built-in functions of R make tables largely unnecessary, as these functions are faster to use and often more accurate than tabled values.

Solution

Finding quantiles for various probability distributions is very easy. You may recall the t distribution has a single parameter: the degrees of freedom. The F distribution has two parameters: the degrees of freedom for the numerator and the degrees of freedom for the denominator. The chi-square distribution has one parameter: the degrees of freedom. Interestingly from a mathematical standpoint, each of these distributions has a theoretical relationship to the normal distribution, though their probability density functions are quite different from that for the normal distribution. The t distribution is symmetrical, like the normal distribution, but the F and chi-square distributions are positively skewed.

Finding p Values for t

Finding one- and two-tailed p values for t is made simple with the pt() function. The pt() function returns the left-tailed, or optionally, the right-tailed cumulative probability for a given value of t and the relevant degrees of freedom. For example, if we have conducted a one-tailed test with 18 degrees of freedom, and the value of t calculated from our sample results was 2.393, the p value is found as follows:

```
> pt(2.393, 18, lower = F)
[1] 0.01391179
```

If the test is two-tailed, we can simply double the one-tailed probability:

```
> 2 * pt(2.393, 18, lower = F)
[1] 0.02782358
```

Finding Critical Values of t

For t tests, we often use two-tailed hypothesis tests, and for such tests, there will be two critical values of t. We also use lower and upper critical values of t to develop confidence intervals for the mean. It is typical practice to use only the right tail of the F and chi-square distributions for hypothesis tests, but in certain circumstances, we use both tails, as in confidence intervals for the variance and standard deviation using the chi-square distribution.

Assume we have the following 25 scores:

```
> x <- rnorm(25, 65, 10)
> x <- round(x, 1)
> x
 [1] 60.2 64.4 64.8 65.6 63.7 68.0 85.7 58.2 53.1 55.0 72.3 50.2 78.5 67.1 79.2
[16] 72.1 59.2 50.6 77.8 56.3 64.0 76.8 68.2 75.0 59.6
```

We can use the t distribution to develop a confidence interval for the population mean. Because we have 25 scores, the degrees of freedom are $n - 1 = 24$. We want to find the values of t that cut off the top 2.5% of the distribution and the lower 2.5%, leaving us with 95% in the middle. Like the standard normal distribution, the t distribution has a mean of zero. Because the t distribution is an estimate of the normal distribution, a confidence interval based on the t distribution will be wider than a confidence interval based on the standard normal distribution. Let us calculate the mean and standard deviation of the data, and then construct a confidence interval. We will use the fBasics package. Begin with the commands install.packages("fBasics") and library(fBasics).

```
> install.packages("fBasics")
> library(fBasics)
> basicStats(x)
                        x
nobs          25.000000
NAs            0.000000
Minimum       50.200000
Maximum       85.700000
1. Quartile   59.200000
3. Quartile   72.300000
Mean          65.824000
Median        64.800000
Sum         1645.600000
SE Mean        1.917057
```

```
LCL Mean      61.867388
UCL Mean      69.780612
Variance      91.877733
Stdev          9.585287
Skewness       0.178608
Kurtosis      -0.968359
```

Note the upper (UCL) and lower (LCL) control limits of the 95% confidence interval for the mean. These are developed using the t distribution. The confidence interval can be defined as:

$$95\% \text{ CI: } M - t_{\alpha/2} \times s_M \leq \mu \leq M + t_{\alpha/2} \times s_M$$

We can use the qt() function to find the critical values of t as follows. As with the normal distribution, the critical values of t are the same in absolute value. The value s_M is the standard error of the mean, which is reported above.

```
> qt(.975, 24)
[1] 2.063899
> qt(.025, 24)
[1] -2.063899
```

Substituting the appropriate values produces the following result, and we see that our confidence limits agree with those produced by the fBasics package.

```
95% CI = 65.824 - 2.063899×1.917057 ≤ µ ≤ 65.824 + 2.063899×1.917057
= 65.824 - 3.956612 ≤ µ ≤  65.824 + 3.956612
= 61.867388 ≤ µ ≤  69.780612
```

The t.test() function in R will automatically produce confidence intervals. For example, if we simply do a one-sample t test with the preceding data, R will calculate the confidence interval for us:

```
> t.test(x)

        One Sample t-test

data:  x
t = 34.336, df = 24, p-value < 2.2e-16
alternative hypothesis: true mean is not equal to 0
95 percent confidence interval:
 61.86739 69.78061
sample estimates:
mean of x
   65.824
```

Finding p Values for F

Unlike the t distribution, the F distribution is positively skewed. It has two degrees of freedom so most textbooks have several pages of tables, which could be eliminated using R. As I mentioned earlier, we usually find critical values of F in the right tail of the distribution. The F ratio is formed from two variance estimates, and it can range from zero to $+\infty$. As a matter of expediency, if you are simply testing the equality of two variance estimates, you should always divide the smaller estimate into the larger one. Though it is possible to test using the left tail of the F distribution, this is rarely done. The pf() function locates the p value and the qf() function gives the quantile for a given F ratio and the relevant degrees of freedom.

For example, assume we have two variance estimates and wish to test the null hypothesis that they are equal against the alternative that they are unequal. Let's create a vector of *y* scores to join with the *x* scores we worked with in Recipe 9-4:

```
> head(df)
     x    y
1 60.2 84.5
2 64.4 89.7
3 64.8 74.0
4 65.6 90.4
5 63.7 84.9
6 68.0 99.6
> tail(df)
      x    y
20 56.3 79.8
21 64.0 82.5
22 76.8 80.7
23 68.2 86.4
24 75.0 81.2
25 59.6 94.1
```

We can use the `sapply()` function to obtain the variances of *x* and *y*.

```
> sapply(df, var)
       x        y
91.87773 43.40677
```

The question is whether the variance for *x*, the larger of the two estimates, is significantly larger than the variance for *y*. There are 25 observations for each variable, so the degrees of freedom for the F ratio are 24 and 24. We supply the value of F, the numerator degrees of freedom, and the denominator degrees of freedom to find the *p* value. To obtain a two-tailed *p* value, we double the one-tailed value:

```
F <- var(x)/var(y)
> F
[1] 2.075825
> pvalue
Error: object 'pvalue' not found
> pvalue <- 2*(pf(F, 24, 24, lower = FALSE)
+ )
> pvalue
[1] 0.07984798
```

We see that the estimates differ by more than a factor of 2, but they are not significantly different at alpha = .05. To compare the means of *x* and *y*, we could therefore use the *t* test that assumes equality of variance. The two means are significantly different, as the *p* value indicates.

```
> t.test(x, y, var.equal = T)

        Two Sample t-test
```

```
data: x and y
t = -8.7317, df = 48, p-value = 1.766e-11
alternative hypothesis: true difference in means is not equal to 0
95 percent confidence interval:
 -25.06795 -15.68405
sample estimates:
mean of x mean of y
   65.824    86.200
```

Finding Critical Values of *F*

Use the qf() function to find quantiles for the *F* distribution. For example, assume we have three groups and each group has 10 observations. For an analysis of variance, the critical value of *F* would be found as follows. Note we have 29 total degrees of freedom, 2 degrees of freedom between groups, and 27 degrees of freedom within groups. We thus need a critical value of *F* for 2 and 27 degrees of freedom that cut off the lower 95% of the *F* distribution from the upper 5%:

```
> qf(0.95, 2, 27)
[1] 3.354131
```

Finding *p* Values for Chi-Square

The chi-square distribution, like the *F* distribution, is positively skewed. It has one parameter, the degrees of freedom. We use the chi-square distribution nonparametrically for tests of model fit (or goodness of fit) and association. The chi-square distribution is also used parametrically for confidence intervals for the variance and standard deviation.

Use the pchisq() function to find the *p* value for a value of chi-square. In an interesting historical side note, Pearson, who developed the chi-square tests of goodness of fit and association, had difficulty with getting the degrees of freedom right. Fisher corrected Pearson—and the two were bitter enemies thereafter. For chi-square used for goodness of fit, the degrees of freedom are not based on sample size, but on the number of categories. We will discuss this in more depth in Chapter 10.

Assume we have a value of chi-square of 16.402 with 2 degrees of freedom. Find the *p* value as follows:

```
> chisq <- 16.402
> pchisq(chisq, 2, lower = FALSE)
[1] 0.0002743791
```

Finding Critical Values of Chi-Square

We can use the qchisq() function for finding critical values of chi-square. For tests of goodness of fit and association (also known as *tests of independence*), we test on the right tail of the distribution. For confidence intervals for the variance and the standard deviation, we find left- and right-tailed values of chi-square.

Now, let us see how to construct a confidence interval for the population variance using the chi-square distribution. We will use the standard 95% confidence interval. We will continue to work with our example data. We found the variance for *x* from our previous table to be 91.87773. A confidence interval for the population variance can be defined as follows:

$$\frac{(n-1)s^2}{\chi^2_{\alpha/2,df}} < \sigma^2 < \frac{(n-1)s^2}{\chi^2_{1-\alpha/2,df}}$$

The right and left values of chi-square cut off the upper and lower 2.5% of the distribution, respectively. Let us find both values. We supply the probability and the degrees of freedom. In this case, the degrees of freedom are $n - 1 = 25 - 1 = 24$.

```
> qchisq(.975, 24)
[1] 39.36408
> qchisq(.025, 24)
[1] 12.40115
```

Now, we can substitute the required values to find the confidence interval for the population variance:

$$\frac{(24)91.87773}{12.40115} < \sigma^2 < \frac{(24)91.87773}{39.36408} = \frac{2205.0655}{12.401158} < \sigma^2 < \frac{2205.0655}{39.3640}$$

$$= 56.017199 < \sigma^2 < 177.81137$$

To find the confidence interval for the standard deviation, simply extract the positive square roots of the upper and lower limits:

$$\sqrt{56.017199} < \sigma < \sqrt{177.81137} = 7.4845 < \sigma < 13.3335$$

■ ■ ■

Hypothesis Tests for Means, Ranks, or Proportions

R provides easy access to virtually any of the traditional hypothesis tests for means, ranks, and proportions one might need. In Chapter 10 you learn how to do hypothesis tests for one sample, for two samples, and for three or more samples. We will include both parametric tests and nonparametric tests, illustrating each with appropriate data. The statistics functions in base R are usually adequate, but it is often helpful to use additional packages for their enhanced features.

Before we delve into hypothesis testing, let us briefly review the scales of measurement and their relationship to data and hypothesis testing. Throughout this book, we have made little distinction between interval and ratio data. The difference, of course, is that ratio data have a true zero point, while the zero point on an interval scale is arbitrary. From a statistical perspective, we combine interval and ratio data into a single category, and we can do parametric statistical analyses with such data as long as the distributional assumptions are warranted. We have also used nominal data, which in the strictest sense are simply categories in which we group objects or individuals with the same attribute. We can use numbers to represent the categories, but a label or some other identifier would work as well. We have a variety of techniques for dealing with nominal data. Between nominal and interval data is the ordinal scale. Ordinal measures give us information about ranks, but they do not tell us about the differences between ranks. Because there are no equal intervals with ordinal data, we must use techniques based on the order of the measures. These techniques are generally called *nonparametric* because they require few, if any, assumptions about population parameters. Moreover, these techniques are not typically used to make estimates or inferences about population parameters.

Recipe 10-1. One-Sample Tests
Problem

Many studies are descriptive in nature and involve the collection of a single sample of data. Often there are no inferences involved in such studies. With descriptive studies, we summarize the data, perhaps graphically as well as numerically, and report our findings. In other studies, however, we are interested in comparing sample values to known or hypothesized population values. In such cases, we use a category of tests known as *one-sample tests*.

Solution

We will discuss one-sample tests for interval and ratio data, and then expand our discussion to include one-sample tests for ordinal and nominal data.

One-Sample z and t Tests for Means

When we describe a sample, we are typically interested in the standard descriptive statistics, including central tendency and variability. You learned to calculate these statistics in Chapter 7. As a quick reminder, we must have interval or ratio data for statistics such as the mean and variance to be meaningful. When we have such data, we can use the fBasics package to get a quick statistical summary. Here again are the summary statistics for the variable x we used in Chapter 9.

```
> require(fBasics)
Loading required package: fBasics
Loading required package: MASS
Loading required package: timeDate
Loading required package: timeSeries
> x
 [1] 60.2 64.4 64.8 65.6 63.7 68.0 85.7 58.2 53.1 55.0 72.3 50.2 78.5 67.1 79.2
[16] 72.1 59.2 50.6 77.8 56.3 64.0 76.8 68.2 75.0 59.6
> basicStats(x)
                        x
nobs            25.000000
NAs              0.000000
Minimum         50.200000
Maximum         85.700000
1. Quartile     59.200000
3. Quartile     72.300000
Mean            65.824000
Median          64.800000
Sum           1645.600000
SE Mean          1.917057
LCL Mean        61.867388
UCL Mean        69.780612
Variance        91.877733
Stdev            9.585287
Skewness         0.178608
Kurtosis        -0.968359
```

In situations where we know the population standard deviation, σ, we can use a z test to compare a sample mean to a known or hypothesized population mean. The one-sample z test is among the simplest of the hypothesis tests, and is typically the first parametric test taught in a statistics course. Assume that we know the population from which the x scores were sampled is normally distributed and has a standard deviation of 11. We want to test the hypothesis that the sample came from a population with a mean of 70. The one-sample z test is not built into the base version of R, but its calculations are relatively simple. We use the following formula, in which the known or hypothesized population mean is subtracted from the sample mean, and the difference is divided by the standard error of the mean.

$$z = \frac{(\bar{x} - \mu)}{\frac{\sigma}{\sqrt{n}}}$$

Assuming we want to do a two-tailed test with alpha = .05, we will use critical values of $z = \pm1.96$. Our simple calculations are as follows:

```
> z <- (mean(x)- 70)/(11/sqrt(25))
> z
[1] -1.898182
```

If you want to create a more effective and reusable z.test function, you can do something like the following. The arguments supplied to the function are the raw data, the test value (mu), the sample size, and the alpha level, which is defaulted to the customary $\alpha = .05$.

```
        }
z.test <- function(x, mu, sigma, n, alpha = 0.05) {
        sampleMean <- mean(x)
        stdError <- sigma/sqrt(n)
        zCrit <- qnorm(1 - alpha/2)
        sampleZ <- (sampleMean-mu)/stdError
        LL <- sampleMean - zCrit*stdError
        UL <- sampleMean + zCrit*stdError
        pValue <- 2 * (1 - pnorm(abs(sampleZ)))
        cat("\t","one-sample z test","\n")
        cat("sample mean:",sampleMean,"\n")
        cat("test value:",mu,"\t","sigma:",sigma,"\n")
        cat("sample z:",sampleZ,"\n")
        cat("p value:",pValue,"\n")
        cat((1-alpha)*100,"percent confidence interval","\n")
        cat("lower:",LL,"\t","upper:",UL,"\n")
}
```

Here is the output from the z.test function shown earlier. For familiarity's sake, I purposely produced output similar to that of the built-in t.test function in R.

```
> z.test(x, mu = 70, sigma = 11, n = 25, alpha = 0.05)
         one-sample z test
sample mean: 65.824
test value: 70     sigma: 11
sample z: -1.898182
p value: 0.05767214
95 percent confidence interval
lower: 61.51208         upper: 70.13592
```

Because you now have a reusable function, you can change the arguments and run the test again. For example, let us change the alpha level to .01 and examine the new confidence interval. Note that as the confidence level increases, the interval width increases correspondingly.

```
> z.test(x,70,11,25,.01)
        One-Sample z Test

sample mean: 65.824
test value: 70    sigma: 11
Sample z: -1.898182
P value: 0.05767214
99 percent confidence interval
Lower: 60.15718        Upper: 71.49082
```

The one-sample and two-sample *z* tests for means are implemented in the BSDA package available through CRAN. We will illustrate the BSDA package in Recipe 10-3.

It is rarely the case that we know the population standard deviation. When the population standard deviation is unknown, we use one-sample *t* tests to test hypotheses about a sample mean. The formula for *t* is identical to that for *z*, but we use the *t* distribution rather than the standard normal distribution to find *p* values and calculate confidence intervals, and we use the sample standard deviation instead of the population standard deviation. The t.test function allows you to test a sample mean against a known or hypothesized population mean when the population standard deviation is unknown. The function performs the *t* test and calculates a 95% confidence interval. The user specifies the data and the hypothesized population mean.

```
> t.test(x, mu = 70)

        One Sample t-test

data:  x
t = -2.1783, df = 24, p-value = 0.03943
alternative hypothesis: true mean is not equal to 70
95 percent confidence interval:
 61.86739 69.78061
sample estimates:
mean of x
   65.824
```

One-Sample Tests for Nominal and Ordinal Data

It is often of interest to determine whether a sample proportion is equal to a known or hypothesized population proportion. For example, a recent Gallup poll showed that 58% of Americans believe higher insurance rates are justified for those who smoke. Assume we conduct the same poll with a random sample in a tobacco-producing state such as North Carolina. We poll 300 adults and determine that 148 of them are in favor of higher insurance rates for smokers. Our sample proportion is as follows:

$$\hat{p} = 148 / 300 = .4933$$

We can use the prop.test function in base R to determine whether the sample proportion of .4933 is significantly different from the hypothesized population proportion of .58:

```
> prop.test(148, 300, .58)

        1-sample proportions test with continuity correction
```

```
data:  148 out of 300, null probability 0.58
X-squared = 8.8978, df = 1, p-value = 0.002855
alternative hypothesis: true p is not equal to 0.58
95 percent confidence interval:
 0.4355593 0.5512809
sample estimates:
        p
0.4933333
```

The results indicate that the sample proportion is significantly lower than the US average.

As a statistical side note, most statistics texts cast the one-sample test of proportion as a z test rather than a chi-square test. The interested reader is referred to a good statistics text for an explanation. A good online source for basic statistics is Dr. David Lane's HyperStat web site at `http://davidmlane.com/hyperstat/`.

'Suffice it to say here that when one calculates z for a one-sample test of proportion, squaring the value of z produces the value of chi-square, and the two tests are statistically and mathematically equivalent. The binom. test function can be used for a proportion, as follows. Note, however, that the prop.test function applies the Yates correction for continuity, so the value of z^2 will not equal chi-square in this case, and the p values and confidence intervals are slightly different for the two tests, though not appreciably so:

```
> binom.test(148, 300, .58)


        Exact binomial test

data:  148 and 300
number of successes = 148, number of trials = 300, p-value = 0.002797
alternative hypothesis: true probability of success is not equal to 0.58
95 percent confidence interval:
 0.4354025 0.5513971
sample estimates:
probability of success
              0.4933333
```

The paired-samples t test can be used to compare two scores for either the same subjects (repeated measures) or matched pairs of scores. The paired-samples t test is a special case of the one-sample t test in that the data of interest are the differences between the paired scores. Similarly, when you have ordinal data, a nonparametric alternative to the one-sample t test is the Wilcoxon signed rank test, which also applies to paired data. The paired-samples t test typically is used to test the hypothesis that the mean difference is zero against the alternative that the mean difference is not zero. Scale data that have been converted to ranks because the data are non-normal or data collected as ranks originally may be analyzed with the Wilcoxon signed rank test.

Assume there is a proposal to build a new school in a residential neighborhood. At the last town council meeting, results of a poll of 13 people in the neighborhood were presented as evidence that there is support for the new school. Unfortunately, the data collection tool was developed by one of the neighborhood residents, and yielded only ordinal data. Here are 13 hypothetical responses (see Table 10-1). Nine of the 13 respondents indicated the school should be built either immediately or soon. There is ordinality to the responses, but they are not on an interval scale. We can transform the responses to ordinal scores from 1 to 5 and test the hypothesis that the average score is 3 (*no opinion*) against the alternative that the median score is not equal to 3 (*no opinion*). We see that resident 8 has no opinion. The customary practice is to discard observations that lead to a zero difference between the score and the test value, so we will create the data frame omitting resident 8.

Table 10-1. *Hypothetical Opinion Poll Results*

Resident	The school is not an asset. Do not build it.	The school may be an asset eventually.	I have no opinion.	The school is an asset. Build it soon.	The school is absolutely necessary. Build it now!
1	X				
2		X			
3				X	
4				X	
5				X	
6					X
7				X	
8			X		
9		X			
10				X	
11					X
12					X
13				X	

The transformed scores are as follows. I included a column for the test value, 3, and the difference between the individual's ranking and the test value for instructive purposes, but these are not used in the test. Instead, one specifies the column of scores and the test value in the arguments to the `wilcox.test` function. Although the test value is technically a hypothetical median score, the function labels this as `mu`.

```
> schoolPoll
   person score testValue diff
1       1     1         3   -2
2       2     2         3   -1
3       3     4         3    1
4       4     4         3    1
5       5     4         3    1
6       6     5         3    2
7       7     4         3    1
9       9     2         3   -1
10     10     4         3    1
11     11     5         3    2
12     12     5         3    2
13     13     4         3    1
```

```
> wilcox.test(schoolPoll$score, mu = 3)

        Wilcoxon signed rank test with continuity correction

data:  schoolPoll$score
V = 58.5, p-value = 0.1217
alternative hypothesis: true location is not equal to 3

Warning message:
In wilcox.test.default(schoolPoll$score, mu = 3) :
  cannot compute exact p-value with ties
```

The test shows that the median score is not significantly different from the test value of 3. For the sake of comparison, the output from SPSS for the Wilcoxon signed rank test for the same data is shown in Figure 10-1. Unsurprisingly, the two programs produce the same result. The inquisitive reader may want to repeat this test using the column of difference scores as the data and with the test value set to zero. The results should be identical to those of the test on the scores themselves.

Hypothesis Test Summary

	Null Hypothesis	Test	Sig.	Decision
1	The median of rank equals 3.000.	One-Sample Wilcoxon Signed Rank Test	.112	Retain the null hypothesis.

Asymptotic significances are displayed. The significance level is .05.

Figure 10-1. *SPSS output for the Wilcoxon signed rank test*

Recipe 10-2. Two-Sample Tests for Related Means, Ranks, and Proportions

Problem

Two-sample tests are many researchers' stock in trade. The independent-samples *t* test is probably the most frequently used hypothesis test in many fields of research. We have discussed the one-sample *t* test and its relationship to the paired-samples *t* test, but let's illustrate it again here. We will also discuss the independent-samples *t* tests, both assuming equal variances in the population and assuming unequal variances. Finally, we will discuss nonparametric tests for two samples using nominal and ordinal data.

Solution

The paired-samples *t* test is a special case of the one-sample *t* test, as we have discussed. Just as we did with the imaginary poll data in Recipe 10-1, we can calculate a column of difference scores for the pairs of observations. Testing the hypothesis that the mean difference is zero shows that there is really just one sample, the difference scores. The degrees of freedom correspond to the number of pairs minus 1, rather than the total number of observations.

With ordinal data, the Wilcoxon signed rank test illustrated in Recipe 10-1 can be used for paired data as well. The McNemar test is used for paired nominal data. As with the 2 × 2 chi-square test for independent samples, the McNemar test is often used in conjunction with the Yates continuity correction.

Let us begin with an example of a paired-samples t test. In this experiment conducted by the author, five-letter stimulus words were flashed briefly on a computer screen to the right or the left of a fixation point in the center of the screen. The duration of the word on the screen increased until the participant could type the correct word. Because of the "wiring" of the human brain, words projected to the left of the fixation point are sent to the right cerebral cortex for processing, whereas those shown on the right of the fixation point go to the left side. On the premise that the left side of the brain is specialized for verbal processing, it should be true that words are recognized more quickly by the left brain than by the right brain. Words sent to the right brain should take longer to be processed because of the extra time required to transfer them to the verbal processing areas on the left side of the brain. The data are as follows:

```
> head(pairedData)
  sex age handed  left right
1   M  24       R 0.115 0.139
2   F  19       R 0.090 0.093
3   F  22       L 0.104 0.133
4   F  18       L 0.101 0.100
5   F  18       L 0.116 0.116
6   F  18       L 0.090 0.113
> tail(pairedData)
   sex age handed  left right
92   F  18       R 0.093 0.096
93   F  19       R 0.106 0.124
94   F  20       R 0.093 0.099
95   F  18       R 0.108 0.108
96   M  18       R 0.110 0.136
97   M  18       R 0.098 0.116
```

For instructive purposes, let us once again calculate a column of difference scores, subtracting the processing time for the left side of the brain from that of the right side of the brain. We will append this column to the data frame by use of the cbind() function. As we suspected, most of the differences are positive, indicating the left side of the brain recognizes words more quickly than the right side does.

```
> diff <- pairedData$right - pairedData$left
> pairedData <- cbind(pairedData, diff)
> head(pairedData)
  sex age handed  left right    diff
1   M  24       R 0.115 0.139  0.024
2   F  19       R 0.090 0.093  0.003
3   F  22       L 0.104 0.133  0.029
4   F  18       L 0.101 0.100 -0.001
5   F  18       L 0.116 0.116  0.000
6   F  18       L 0.090 0.113  0.023
```

Now, let us perform a paired-samples t test followed by a one-sample t test on the difference scores testing the hypothesis that the mean difference is zero against the two-sided alternative that the mean difference is not zero. Note that apart from slight labelling differences, the two tests produce the same results. The left side of the brain is significantly faster than the right side at recognizing words.

```
> t.test(pairedData$right, pairedData$left, paired = T)

        Paired t-test

data:  pairedData$right and pairedData$left
t = 7.6137, df = 96, p-value = 1.85e-11
alternative hypothesis: true difference in means is not equal to 0
95 percent confidence interval:
 0.007362381 0.012555144
sample estimates:
mean of the differences
            0.009958763


> t.test(pairedData$diff, mu = 0)

        One Sample t-test

data:  pairedData$diff
t = 7.6137, df = 96, p-value = 1.85e-11
alternative hypothesis: true mean is not equal to 0
95 percent confidence interval:
 0.007362381 0.012555144
sample estimates:
  mean of x
0.009958763
```

The data for both left- and right-brain word recognition are positively skewed. The difference data, however, are more normally distributed, as the histograms and normal q-q plot indicate (see Figure 10-2). Remember you can set the graphic device to show multiple graphs using the par function. We display histograms for the left- and right-brain word recognition data and the difference data, as well as a normal q-q plot for the difference data.

```
> par(mfrow=c(2,2))
> hist(pairedData$left)
> hist(pairedData$right)
> hist(pairedData$diff)
> qqnorm(pairedData$diff)
```

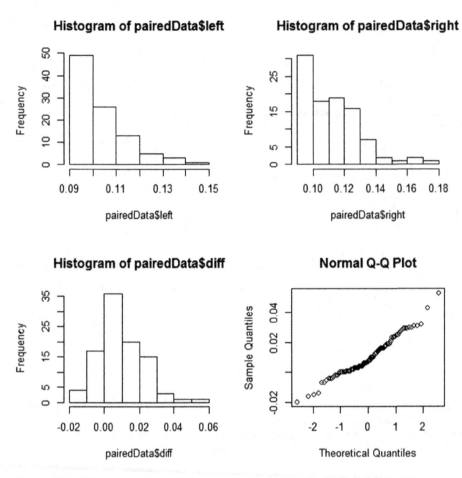

Figure 10-2. *Histograms of the word recognition data*

We can use the same data to illustrate the Wilcoxon signed rank test for paired data. The normal q-q plot shown in Figure 10-2 indicates a roughly linear relationship between the observed and theoretical quantiles, but the histogram shows the difference data are still slightly positively skewed and that there are outliers in the data. We can convert the data to ranks and then perform the signed rank test to determine if the average difference has a "location shift" of zero. This is the rough equivalent of testing the hypothesis that the median difference score is zero. Our paired-samples *t* test has already indicated that the left side of the brain is significantly faster at word recognition than the right side. Let us determine whether the difference is still significant when we use the nonparametric test. The signed rank test is significant.

```
> wilcox.test(pairedData$right, pairedData$left, paired = T)

        Wilcoxon signed rank test with continuity correction

data:  pairedData$right and pairedData$left
V = 3718, p-value = 1.265e-10
alternative hypothesis: true location shift is not equal to 0
```

For nominal data, we can use the McNemar test for nominal data representing repeated measurements of the same objects or individuals or matched pairs of subjects. This test applies to 2 × 2 contingency tables in which the data are paired. The table is constructed as follows (see Table 10-2).

Table 10-2. *Table Layout for the McNemar Test*

	Measure 2 Positive	Measure 2 Negative	Row Total
Measure 1 Positive	a	B	$a + b$
Measure 1 Negative	c	D	$c + d$
Column Total	$a + c$	$b + d$	n

The test statistic is chi-square with 1 degree of freedom, calculated as follows:

$$x^2 = \frac{(b-c)^2}{b+c}$$

Note that the main diagonal of the table is not used to calculate the value of chi-square. Instead, the off-diagonal entries are used. We are testing the hypothesis that the off-diagonal entries b and c are equal. As with the 2 × 2 contingency table for independent samples, the Yates continuity correction may be applied, in which case the value of chi-square is computed as follows:

$$\chi^2 = \frac{(|b-c|-0.5)^2}{b+c}$$

As a concrete example, assume we asked a sample of 1,009 people whether they approved or disapproved of George W. Bush's handling of his job as president immediately before and immediately after the September 11 attacks. The hypothetical data[1] are shown in Table 10-3.

Table 10-3. *Hypothetical Data for McNemar Test*

	After Approve	After Disapprove	Row Total
Before Approve	461	53	514
Before Disapprove	443	52	495
Column Total	904	105	1009

```
> BushApproval <- matrix(c(461,53,443,52),ncol = 2)
> BushApproval
     [,1] [,2]
[1,]  461  443
[2,]   53   52
> mcnemar.test(BushApproval)
```

---

[1]These data are loosely based on the Gallup poll results reported at http://www.gallup.com/poll/116500/presidential-approval-ratings-george-bush.aspx.

```
        McNemar's Chi-squared test with continuity correction

data:  BushApproval
McNemar's chi-squared = 305.0827, df = 1, p-value < 2.2e-16
```

We reach the rather obvious conclusion (which was supported by the Gallup polls in 2001) that Mr. Bush's approval ratings increased significantly after 9/11.

The McNemar test is effective for tables in which $b + c > 25$. When $b + c < 25$, a binomial test can be used to determine the exact probability that $b = c$.

Recipe 10-3. Two-Sample Tests for Independent Means, Ranks, and Proportions

Problem

The use of independent samples to determine the effectiveness of some treatment or manipulation is common in research. The independent-samples t test applies when the dependent variable is interval or ratio in nature. We can use either the t test assuming equal variances in the population or the test assuming unequal variances. With ordinal data, the nonparametric alternative to the t test is the Mann-Whitney U test. We can use chi-square or z tests to compare independent proportions.

Solution

Parametric tests such as the t test and the analysis of variance (ANOVA) assume the data are interval or ratio, that the data are sampled from a normally distributed population, and that the variances of the groups being compared are equal in the population. To the extent that these assumptions are met, the standard hypothesis tests are effective. The central limit theorem allows us to relax the assumption of normality of the population distribution, as the sampling distribution of means converges on a normal distribution as the sample size increases, regardless of the shape of the parent population.

The traditional alternative to parametric tests when distributional assumptions are violated, as we have discussed and illustrated, is the use of distribution-free or nonparametric tests. However, the independent-samples t test assuming unequal variances is often a preferable alternative. In Chapter 12, you will learn additional alternatives, most of which have been developed in the past 50 years, for dealing with data that are not normally distributed.

A two-sample z test is applicable when the population standard deviations are known or when the samples are large. As with the one-sample test, the two-sample z test is not built into base R. As mentioned previously, both one-sample and two-sample z tests are available in the BSDA package. The z.test function in BSDA is quite flexible. Here is the one-sample z test output for our x array, and the results are identical to those from the z.test function we wrote earlier in Recipe 10-1.

```
> install.packages("BSDA")
> library(BSDA)
Loading required package: e1071
Loading required package: lattice

Attaching package: 'BSDA'

The following object is masked from 'package:datasets':

    Orange
```

```
> z.test(x, mu = 70, sigma.x = 11)

        One-sample z-Test

data:  x
z = -1.8982, p-value = 0.05767
alternative hypothesis: true mean is not equal to 70
95 percent confidence interval:
 61.51208 70.13592
sample estimates:
mean of x
   65.824
```

A two-sample z test can be conducted using either known (unlikely) values for the population standard deviations or using the sample standard deviations (more likely) as estimates of the population parameters. The two-sample z test will produce identical test statistics as those of the independent-samples t test when the sample standard deviations are used, as the following output illustrates. See that the Welch t test adjusts the degrees of freedom to compensate for unequal variances. The data for x and y are as follows. Assume they represent the hypothetical scores of two different classes on a statistics course final.

```
> x
 [1] 60.2 64.4 64.8 65.6 63.7 68.0 85.7 58.2 53.1 55.0 72.3 50.2 78.5 67.1 79.2
[16] 72.1 59.2 50.6 77.8 56.3 64.0 76.8 68.2 75.0 59.6
> y
 [1] 84.5 89.7 74.0 90.4 84.9 99.6 90.0 94.2 91.2 95.2 91.4 79.0 75.9 77.9 90.9
[16] 79.7 85.6 85.2 91.0 79.8 82.5 80.7 86.4 81.2 94.1

> z.test(x, y, sigma.x = sd(x), sigma.y = sd(y))

        Two-sample z-Test

data:  x and y
z = -8.7317, p-value < 2.2e-16
alternative hypothesis: true difference in means is not equal to 0
95 percent confidence interval:
 -24.94971 -15.80229
sample estimates:
mean of x mean of y
   65.824    86.200


> t.test(x, y)

      Welch Two Sample t-test

data:  x and y
t = -8.7317, df = 42.768, p-value = 4.686e-11
alternative hypothesis: true difference in means is not equal to 0
95 percent confidence interval:
 -25.08283 -15.66917
sample estimates:
mean of x mean of y
   65.824    86.200
```

When the assumption of homogeneity of variance is justified, the independent-samples test assuming equal variances in the population is slightly more powerful than the Welch test. Let us compare the two tests. To conduct the pooled-variance t test, set the var.equal argument to T or TRUE. The value of t is the same for both tests, but the p values, degrees of freedom, and confidence intervals are slightly different. For the pooled-variance t test, the degrees of freedom are $n_1 + n_2 - 2$.

```
> t.test(x, y, var.equal = T)

        Two Sample t-test

data:  x and y
t = -8.7317, df = 48, p-value = 1.766e-11
alternative hypothesis: true difference in means is not equal to 0
95 percent confidence interval:
 -25.06795 -15.68405
sample estimates:
mean of x mean of y
   65.824    86.200
```

The wilcox.test function can be used to conduct an independent-groups nonparametric test. This test is commonly known and taught as the Mann-Whitney U test, but technically would be better labelled the Wilcoxon-Mann-Whitney sum of ranks test. To conduct the Mann-Whitney test, use the following syntax. We will continue with our example of the x and y vectors we worked with earlier. The Mann-Whitney test is often described as a test comparing the medians of the two groups, but it is really testing the locations of the two distributions. We see that when the data are converted to ranks, the difference is still significant.

```
> wilcox.test(x, y)

        Wilcoxon rank sum test

data:  x and y
W = 33, p-value = 8.5e-10
alternative hypothesis: true location shift is not equal to 0
```

Comparing two independent proportions can be accomplished by a z test or by a chi-square test, as we have discussed previously. Let us compare the two sample proportions $\hat{p}_1 = x_1 / n_1$ and $\hat{p}_2 = x_2 / n_2$ where x_i and n_i represents the counts of the occurrences and the samples size for sample i. A two-sample z test for proportions is calculated as follows:

$$z = \frac{\hat{p}_1 - \hat{p}_2}{\sqrt{\dfrac{\overline{pq}}{n_1} + \dfrac{\overline{pq}}{n_2}}}$$

where $\overline{p}$ is the pooled sample proportion, $(x_1 + x_2) / (n_1 + n_2)$, and $\overline{q} = 1 - \overline{p}$.

As we have discussed, the square of the z statistic calculated by use of the preceding formula is distributed as chi-square with 1 degree of freedom. Let us do a one-tailed test with the following data to determine if airbags in automobiles save lives. The hypothetical data are as follows (see Table 10-4).

Table 10-4. *Hypothetical Airbag Data*

	Airbag Present	No Airbag Present	Row Total
Occupant or Driver Fatality	41	58	99
Total Number of Occupants	11548	9857	21405
Column Total	11589	9915	21504

Either the chisq.test or the prop.test function can be used to test the hypothesis that the proportions are equal against the alternative that they are unequal. First, however, for instructive purposes, let us calculate the value of z using the formula shown earlier. We can write a reusable function as follows, passing the arguments x_1, x_2, n_1, and n_2, which are the number of "successes" and the column totals in Table 10-4.

```
z.prop <- function(x1, x2, n1, n2) {
      p1 <- x1/n1
      p2 <- x2/n2
      p_bar <- (x1+x2)/(n1+n2)
      q_bar <- 1-p_bar
      pq <- p_bar * q_bar
      z <- (p1 - p2)/sqrt((pq/n1)+(pq/n2))
      return(z)
      }
> z.prop(41,58,11589,9915)
[1] -2.496433
```

Now, for the sake of comparison, run the same test using the prop.test function in R. See that the square of z obtained by our function equals the uncorrected value of chi-square from the prop.test function.

```
> prop.test(x = c(41,58), n = c(11589, 9915), correct = FALSE)

        2-sample test for equality of proportions without continuity
        correction

data:  c(41, 58) out of c(11589, 9915)
X-squared = 6.2322, df = 1, p-value = 0.01254
alternative hypothesis: two.sided
95 percent confidence interval:
 -0.0041616709 -0.0004620991
sample estimates:
     prop 1      prop 2
0.003537838 0.005849723
```

Recipe 10-4. Tests for Three or More Means

Problem

In many research studies, there are three or more means to compare. Although it is tempting to perform multiple t tests to test for mean differences, this results in a compounding Type I error. As a result, we are much more likely to reject a true null hypothesis than we would be for a single test at the same nominal alpha level. If each of the c tests is conducted with the same nominal alpha level, the actual Type I error rate compounds as follows:

$$\alpha_{tot} = 1 - (1 - \alpha)^c$$

The analysis of variance (ANOVA) controls the overall rate of a Type I error to no more than alpha, allowing us to determine if there are mean differences among two or more samples without inflating the probability of a Type I error.

Solution

To illustrate, with six means, there are $_6C_2 = 15$ pairwise comparisons possible. Thus, the overall Type I error rate would be $1 - .95^{15} = .54$, which is clearly far too high. R. A. Fisher developed the ANOVA to perform a single overall test with a controlled Type I error rate. If the overall test is significant, at least one pair of means is significantly different, and depending on the circumstances, one might conduct post hoc comparisons at a controlled Type I error rate to determine which means differ.

The one-way ANOVA compares means from independent samples. This test is a direct extension of the independent-samples t test, and produces identical results to the t test when two samples are compared. The repeated-measures ANOVA compares two or more means for the same subjects under different conditions or at different points in time. The repeated-measures ANOVA is a direct extension of the paired-samples t test. It is also possible to have multiple factors in an ANOVA. We will discuss the simplest of these cases, a balanced factorial design with two factors.

The parametric ANOVA assumes that the data are at least interval, that the measurements within a group are independent, and that the variances of the groups are equal in the population. When these assumptions are violated, the nonparametric alternatives to the one-way ANOVA and repeated-measures ANOVA are the Kruskal-Wallis analysis of variance for ranks and the Friedman test, respectively.

The aov and anova functions are part of base R. The ez package provides the ezANOVA function, which makes it easier to do between-groups and within-subjects ANOVAs, as well as mixed-model designs. Let us start with the one-way case. This is called a *between-groups design* because the groups are independent. As with the independent-samples t test, there is no requirement that the sample sizes must be equal for the one-way ANOVA. The overall F statistic is calculated as the ratio of the between-groups variance to the within-groups (error) variance. If the null hypothesis that the means are equal is true, the F ratio will be very close to 1. As the between-groups (treatment) variance becomes larger, the F ratio increases in value. We will begin our discussion with one-way tests for means and ranks.

One-Way Tests

The following data are from a study conducted at the University of Melbourne. The pain tolerance threshold was calculated for adult men and women with different hair colors. The higher the pain number, the greater the person's tolerance for pain. Side-by-side boxplots are helpful to visualize the data. We see an obvious connection between hair color and pain thresholds. People with light blond hair are more pain tolerant than others. We will test the significance of this effect with a one-way ANOVA. The qplot function in ggplot2 produces the side-by-side boxplots (see Figure 10-3).

```
> hair
          color pain
1     LightBlond   62
2     LightBlond   60
3     LightBlond   71
4     LightBlond   55
5     LightBlond   48
6     DarkBlond    63
7     DarkBlond    57
8     DarkBlond    52
9     DarkBlond    41
10    DarkBlond    43
11 LightBrunette   42
12 LightBrunette   50
13 LightBrunette   41
14 LightBrunette   37
15  DarkBrunette   32
16  DarkBrunette   39
17  DarkBrunette   51
18  DarkBrunette   30
19  DarkBrunette   35

> install.packages("ggplot2")
> library(ggplot2)
> qplot(factor(color), pain, data = hair, geom = "boxplot")
```

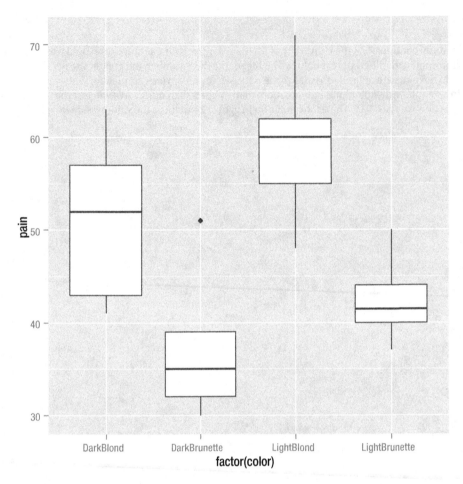

Figure 10-3. *Pain threshold by hair color*

Using the aov function, we see that the pain thresholds are statistically significantly different, though as yet we are not sure which means differ from one another. A popular post hoc test to compare pairs of means is the Tukey HSD (honestly significant difference) test, which is built into the base version of R as the function TukeyHSD. There are many other post hoc procedures, including Bonferroni corrections, the Scheffe test, and the Fisher LSD (least significant difference) criterion. First, examine the ANOVA summary table, and then the results of the Tukey HSD test, which shows that people with light blond hair have significantly higher pain thresholds than light and dark brunettes, but that light and dark blondes do not differ significantly.

```
> summary(model <- aov(pain ~ color, data = hair))
            Df Sum Sq Mean Sq F value  Pr(>F)
color        3   1361   453.6   6.791 0.00411 **
Residuals   15   1002    66.8
---
Signif. codes:  0 '***' 0.001 '**' 0.01 '*' 0.05 '.' 0.1 ' ' 1
> TukeyHSD(model)
  Tukey multiple comparisons of means
    95% family-wise confidence level
```

```
Fit: aov(formula = pain ~ color, data = hair)

$color
                            diff        lwr         upr       p adj
DarkBrunette-DarkBlond      -13.8 -28.696741   1.0967407 0.0740679
LightBlond-DarkBlond          8.0  -6.896741  22.8967407 0.4355768
LightBrunette-DarkBlond      -8.7 -24.500380   7.1003795 0.4147283
LightBlond-DarkBrunette      21.8   6.903259  36.6967407 0.0037079
LightBrunette-DarkBrunette    5.1 -10.700380  20.9003795 0.7893211
LightBrunette-LightBlond    -16.7 -32.500380  -0.8996205 0.0366467
```

The pain data do not appear to have equal variances, as the boxplots in Figure 10-3 imply. The car (Companion to Applied Regression) package provides several tests of homogeneity of variance, including the Levene test, which is used by SPSS as well. Let us determine whether the variances differ significantly. First, we will use the aggregate function to show the variances for the four groups, and then we will conduct the Levene test. The lack of significance for the Levene test is a function of the small sample sizes.

```
> aggregate(pain ~ color, data = hair, var)
         color     pain
1     DarkBlond 86.20000
2   DarkBrunette 69.30000
3    LightBlond 72.70000
4  LightBrunette 29.66667
> > install.packages("car")
> library(car)
> leveneTest(pain ~ color, data = hair)
Levene's Test for Homogeneity of Variance (center = median)
      Df F value Pr(>F)
group  3  0.3927   0.76
      15
```

Just as the independent-samples t test provides the options for assuming equal or unequal variances, the oneway.test function in R provides a Welch-adjusted F test that does not assume equality of variance. The results are similar to those of the standard ANOVA, which assumes equality of variance, but as with the t test, the values of p and the degrees of freedom are different for the two tests.

```
> oneway.test(pain ~ color, data = hair)

        One-way analysis of means (not assuming equal variances)

data:  pain and color
F = 5.8901, num df = 3.00, denom df = 8.33, p-value = 0.01881
```

When the assumptions of normality or equality of variance are violated, we can perform a Kruskal-Wallis analysis of variance by ranks using the kruskal.test function in base R. This test makes no assumptions of equality of variance or normality of distribution. It determines whether the distributions have the same shape, and is not a test of the equality of medians, though some statistics authors claim it is. The test statistic for the Kruskal-Wallis test is called H, and it is distributed as chi-square when the null hypothesis is true.

```
> kruskal.test(hair$pain, hair$color)

        Kruskal-Wallis rank sum test

data:  hair$pain and hair$color
Kruskal-Wallis chi-squared = 10.5886, df = 3, p-value = 0.01417
```

Two-Way Tests for Means

The two-way ANOVA adds a second factor, so that the design now incorporates a dependent variable and two factors, each of which has at least two levels. Thus the most basic two-way design will have four group means to compare. We will consider only the simple case of a balanced factorial design with equal numbers of observations per cell. With the two-way ANOVA, you can test for main effects for each of the factors and for the possibility of an interaction between them. From a research perspective, two-way ANOVA is efficient, because each subject is exposed to a combination of treatments (factors). Additionally, interactions are often of interest because they show us the limits or boundaries of the treatment conditions. Although higher-order and mixed-model ANOVAs are also possible, the interpretation of the results becomes more difficult as the design becomes more complex.

The two-way ANOVA can be conceptualized as a table with rows and columns. Let the rows represent Factor A, and the columns represent Factor B. In a balanced factorial design, each cell represents an independent group, and all groups have equal sample sizes. Post hoc comparisons are not necessary or possible when the design is a 2 × 2 factorial ANOVA, so we will illustrate with a 2 × 3 example. We consider only the fixed effects model here, but the interested reader should know that there are also random effects models in which the factor levels are assumed to be samples of some larger population rather than fixed in advance of the hypothesis tests. With the two-way ANOVA, the total sum of squares is partitioned into the effects due to Factor A, those due to Factor B, those due to the interaction of Factors A and B, and an error or residual term. Thus we have three null hypotheses in the two-way ANOVA, one for each main effect, and one for the interaction.

Assume we have data representing the job satisfaction scores for junior accountants after one year on their first job. The factors are investment in the job (high, medium, or low) and quality of alternatives (high or low). High quality alternatives represent attractive job opportunities with other employers. Although these data are fabricated, they are consistent with those presented by Rusbult and Farrell, the creators of the investment model,[2] which has been validated both in the workplace and in the context of romantic relationships. Assume we have a total of 60 subjects, so the two-way table has 10 observations per cell. For this example, let's use the ezANOVA function in the ez package. The data are as follows:

```
> head(jobSat)
  subject satisf alternatives investment
1       1     52          low        low
2       2     42          low        low
3       3     63          low        low
4       4     48          low        low
5       5     51          low        low
6       6     55          low        low
```

---

[2] See Rusbult, C. E., and Farrell, D. (1983). A longitudinal test of the investment model. The impact on job satisfaction, job commitment, and turnover of variations in reward, costs, alternatives, and investments. *Journal of Applied Psychology, 68,* 429–438.

```
> tail(jobSat)
   subject satisf alternatives investment
55      55     87         high         high
56      56     80         high         high
57      57     88         high         high
58      58     79         high         high
59      59     83         high         high
60      60     80         high         high
```

The output from the ezANOVA function shows that there is a significant main effect for investment and a significant interaction, but that the main effect for alternatives is not significant. When the interaction is significant, it is advisable to plot it in order to understand and interpret it before examining the main effects.

```
> install.packages("ez")
> library(ez)
> ezANOVA(data = jobSat, dv = satisf, wid = subject, between = .(alternatives, investment))
Warning: Converting "subject" to factor for ANOVA.
$ANOVA
                   Effect DFn DFd           F          p p<.05        ges
1              alternatives   1  54   0.8413829 3.630803e-01          0.01534212
2                investment   2  54 307.6647028 3.036366e-30      *   0.91932224
3 alternatives:investment   2  54  29.4058656 2.296500e-09      *   0.52132638

$`Levene's Test for Homogeneity of Variance`
  DFn DFd      SSn    SSd        F          p p<.05
  1 5   54 68.13333 519.3 1.416984 0.2329502
```

The Levene test reveals that the variances can be assumed to be homogeneous. The interaction plot can be generated by use of the interaction.plot function (see Figure 10-4).

```
> with(jobSat, interaction.plot(alternatives, investment, satisf, mean))
```

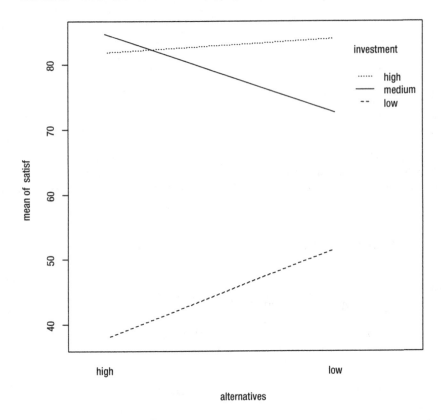

Figure 10-4. *Interaction plot*

Cautious interpretation of the main effect for investment indicates that for those who have made medium or high investment, job satisfaction is higher than for those who have made lower investment. However, the interaction effect shows that job satisfaction is lower for those who have medium investment if they have fewer job alternatives.

A little-known nonparametric alternative to the two-way ANOVA is the Scheirer-Ray-Hare (SRH) test. This test is not currently available in base or any contributed package I am able to locate. The SRH test is a direct extension of the Kruskal-Wallis analysis of variance by ranks to the two-way case. Not all statisticians agree that the SRH test is appropriate for ranked data. For example, Dr. Larry Toothaker of the University of Oklahoma discourages its use because it inflates the Type I error rate and does not test the interaction term correctly. A more effective nonparametric alternative to the factorial ANOVA is the aligned rank transform (ART) test. This test applies the traditional factorial ANOVA after the data have been "aligned" and ranked. A Windows program developed by Dr. Jacob Wobbrock of the University of Washington for aligning and ranking data can be found at `http://depts.washington.edu/aimgroup/proj/art/`.

Recipe 10-5. Repeated-Measures Designs
Problem

Repeated-measures designs involve measures of the same subjects either at different points in time or under different conditions. These are also known as *within-subjects designs*. One can test the subject effect and the treatment effect, though the subject effect is commonly not tested. In this design, each subject serves as its own control, and individual differences are allocated to systematic variation rather than to error. As a result, the error variation is reduced accordingly, and the tests of treatment effects are generally more powerful in repeated-measures designs than in between-groups designs.

Solution

We will consider the repeated measures design and illustrate its power relative to the one-way between-groups ANOVA. Then we will discuss the Friedman test as a nonparametric alternative to the repeated-measures ANOVA.

The repeated-measures design can be thought of as a special case of the two-way ANOVA with one observation per cell. With one observation, there is no within-cell variance, but the cells for a single subject can be compared to the overall subject total. The sums of squares for treatments and for subjects are calculated as in the one-way ANOVA. The absence of within-cell variation means there is no ability to test an interaction effect, and the subject x treatment interaction mean square is used as the error term in repeated-measures ANOVA. The repeated-measures ANOVA also applies to experiments in which randomized blocks are used to control for "nuisance" factors.

To illustrate the power of the repeated-measures design relative to the between-groups design, let us consider the following example modified from Welkowitz, Cohen, and Ewen (2006). In this hypothetical example, subjects listened to classical music while performing a spatial ability test. The study was designed to test the so-called "Mozart" effect in which listening to classical music—especially that of Mozart—is believed to improve spatial reasoning, at least momentarily. We will first treat the data as a between-groups design and then run a repeated-measures analysis assuming that each subject listened to all five composers. We see that the one-way ANOVA is not significant:

```
> music <- read.csv("music.csv")
> head(music)
  Person Subject Composer Score
1      1       1   Mozart    16
2      2       2   Mozart    16
3      3       3   Mozart    14
4      4       4   Mozart    13
5      5       5   Mozart    12
6      6       1   Chopin    16
> tail(music)
   Person Subject  Composer Score
20     20       5  Schubert    12
21     21       1 Beethoven    14
22     22       2 Beethoven    13
23     23       3 Beethoven    13
24     24       4 Beethoven    10
25     25       5 Beethoven    10
> factor(music$Composer)
 [1] Mozart    Mozart    Mozart    Mozart    Mozart    Chopin    Chopin
 [8] Chopin    Chopin    Chopin    Bach      Bach      Bach      Bach
[15] Bach      Schubert  Schubert  Schubert  Schubert  Schubert  Beethoven
[22] Beethoven Beethoven Beethoven Beethoven
Levels: Bach Beethoven Chopin Mozart Schubert
> between <- lm(Score ~ Composer, data = music)
> anova(between)
Analysis of Variance Table
```

```
Response: Score
          Df Sum Sq Mean Sq F value Pr(>F)
Composer   4   14.4    3.60  0.8411 0.5154
Residuals 20   85.6    4.28
```

Now, let us recast the analysis as a repeated-measures ANOVA. We assume each person listened to all five composers and completed five parallel versions of the spatial ability test. To control for order and sequencing effects, let us further assume that each person listened to the five composers in random order, controlled by the computer. Although there are many different ways to run the repeated-measures ANOVA, one effective way is to use the ezANOVA function in the ez package.

```
> ezANOVA(data = music, dv = Score, wid = Subject, within = Composer)
Warning: Converting "Subject" to factor for ANOVA.
$ANOVA
    Effect DFn DFd       F         p p<.05   ges
2 Composer   4  16 5.538462 0.005399203   * 0.144

$`Mauchly's Test for Sphericity`
    Effect        W          p p<.05
2 Composer 0.591716 0.9989984

$`Sphericity Corrections`
    Effect   GGe      p[GG] p[GG]<.05      HFe       p[HF] p[HF]<.05
2 Composer 0.845 0.009177744         * 6.008065 0.005399203         *
```

The F ratio for the repeated-measures analysis is significant at $p = .005$. The repeated-measures ANOVA assumes sphericity, which is the assumption that the differences between the variances of the repeated measures are equal in the population. Unlike the homogeneity of variance assumption in one-way ANOVA, the sphericity assumption is more restrictive, and violations of it lead to substantial increases in Type I error. The Mauchly test confirms that we can assume the sphericity assumption was met in this case. In cases where the sphericity assumption is violated, the two common corrections are to multiply the degrees of freedom by a correction factor known as ε (Greek letter epsilon). The first method of calculating ε is known as the Greenhouse-Geisser (GGe), and the second is known as the Huyn-Feldt (HFe).

When the overall F ratio is significant, it is customary to employ post hoc tests to determine the patterns of significance in the repeated measures. The two most common approaches are to use the Tukey HSD test illustrated in Recipe 10-4 or to use Bonferroni-adjusted comparisons. Let us illustrate the use of the Bonferroni technique. Because there are $_5C_2 = 10$ pairs of means, the adjusted alpha level for the Bonferroni comparisons would be $.05 / 10 = .005$. The Bonferroni-adjusted comparisons reveal that only the comparison between Beethoven and Mozart is significant. Here are the means followed by the Bonferroni-adjusted pairwise t tests.

```
> aggregate(Score ~ Composer, data = music, mean)
   Composer Score
1      Bach  12.4
2 Beethoven  12.0
3    Chopin  13.2
4    Mozart  14.2
5  Schubert  13.2
```

```
> with(music, pairwise.t.test(Score, Composer, p.adjust.method = "bonferroni", paired = T))

        Pairwise comparisons using paired t tests

data:  Score and Composer

          Bach  Beethoven Chopin Mozart
Beethoven 1.000 -         -      -
Chopin    1.000 1.000     -      -
Mozart    0.213 0.042     0.890  -
Schubert  1.000 1.000     1.000  0.890

P value adjustment method: bonferroni
```

The Friedman test is a nonparametric alternative to the repeated-measures ANOVA. As with most other nonparametric tests, the dependent variable is converted to ranks. In the case of repeated-measures, the data are ranked for a given subject, and the ranks for the treatments across subjects are tested for significance. When the overall Friedman test is significant, the most common follow-up test is the use of multiple Wilcoxon signed rank tests. As with the parametric tests, it is important to recognize and control for the inflated Type I error resulting from multiple pairwise comparisons. Here is the Friedman test applied to the current data. For the sake of convenience, I transformed the data to a "wide" format and saved it as a matrix to simplify the application of the Friedman test. The test shows that there is an effect of the composer, just as the parametric test indicated.

```
> musicWide
  Mozart Chopin Bach Schubert Beethoven
1     16     16   16       16        14
2     16     14   14       14        13
3     14     13   12       12        13
4     13     13   10       12        10
5     12     10   10       12        10
> musicWide <- as.matrix(musicWide)
> friedman.test(musicWide)

        Friedman rank sum test

data:  musicWide
Friedman chi-squared = 11.1169, df = 4, p-value = 0.02528
```

CHAPTER 11

■ ■ ■

Relationships Between and Among Variables

In Chapter 11, you learn how to examine and test the significance of relationships between two or more variables measured at the scale (interval and ratio), ordinal, and nominal levels. The study of relationships is an important part of statistics, because we use correlational studies to answer important research questions in a variety of disciplines.

We begin with correlation and regression, examining relationships between a single predictor and a single criterion or dependent variable. We then move to special cases of the correlation coefficient that involve ranked or nominal data for one or both of the variables. Next, we expand the discussion to include a brief introduction to multiple correlation and regression, and we show the importance of this very general technique by relating it to the hypothesis tests we studied in Chapter 10.

Recipe 11-1. Determining Whether Two Scale Variables Are Correlated

Problem

Francis Galton developed the concept of correlation, but left his protégé Karl Pearson to work out the formula. We find that many variables are related to each other in a linear fashion, but correlation does not prove causation. For example, Galton measured various characteristics of more than 9,000 people, including their head circumference. His belief was that people with larger heads must have larger brains and that people with larger brains must be more intelligent. That hypothesis was refuted for years, but recent analyses have shown that indeed there is a small but significant relationship between brain volume and intelligence. A study of more than 1,500 people in 37 samples produced an estimate of the population correlation between the volume of the brain measured in vivo by MRI and an intelligence of .33. Correlation and regression are used in many disciplines to model linear relationships between two or more variables. We begin with the bivariate case for scale data and then study variations on the correlation coefficient for examining relationships when the variables are not measured on a scale.

The correlation coefficient can be understood best by first understanding the concept of *covariation*. The covariance of x (the predictor) and y (the criterion) is the average of the cross-products of their respective deviation scores for each pair of data points. The covariance, which we will designate as σ_{xy}, can be positive, zero, or negative. In other words, the two variables may be positively related (they both go up together and down together); not at all related; or negatively related (one variable goes up while the other goes down). The population covariance is defined as follows:

$$\sigma_{xy} = \frac{\sum(x - \mu_x)(y - \mu_y)}{N}.$$

where N is the number of pairs of (x, y) data points. When we do not know the population mean, we substitute the sample mean and divide the sum of the cross-products of the deviation scores by $n - 1$ as a correction factor:

$$s_{xy} = \frac{\sum (x - \bar{x})(y - \bar{y})}{n - 1}.$$

The covariance is a useful index, but it is affected by the units of measurement for x and y. The covariance of football players' heights and weights expressed in centimeters and kilograms is different from the covariance expressed in inches and pounds. The correlation coefficient corrects this by dividing the covariance by the product of the standard deviations of x and y, resulting in an index that ranges from –1 through 0 to +1, representing a perfect negative relationship, no linear relationship, and a perfect positive relationship. The correlation between the players' heights and weights is invariant whether the height units are inches or centimeters, and whether the weights are expressed in pounds or kilograms. In the population, we define the correlation coefficient as ρ (the Greek letter *rho*):

$$\rho_{xy} = \frac{\sigma_{xy}}{\sigma_x \sigma_y}.$$

In the sample, we use the sample statistics as estimates of the population parameters.

Solution

Let us illustrate the "scalelessness" property of the correlation coefficient by working out the example used earlier. The following data represent the 2014 roster for the New Orleans Saints. Heights are recorded in inches and centimeters, and weights are recorded in pounds and kilograms.

```
> saints <- read.csv("saintsRoster.csv")
> head(saints)
    Number            Name Position Inches Pounds centimeters kilograms Age
1       72 Armstead, Terron       T     77    304      195.58    138.18  23
2       61  Armstrong, Matt       C     74    302      187.96    137.27  24
3       27    Bailey, Champ      CB     73    192      185.42     87.27  36
4       36     Ball, Marcus       S     73    209      185.42     95.00  27
5        9      Brees, Drew      QB     72    209      182.88     95.00  35
6       34  Brooks, Derrius      CB     70    192      177.80     87.27  26
  Experience          College
1          1 Arkansas-Pine Bluff
2          R  Grand Valley State
3         16             Georgia
4          1             Memphis
5         14              Purdue
6          1    Western Kentucky
> tail(saints)
    Number            Name Position Inches Pounds centimeters kilograms Age
70      75   Walker, Tyrunn       DE     75    294      190.50    133.64  24
71      42   Warren, Pierre        S     74    200      187.96     90.91  22
72      82 Watson, Benjamin       TE     75    255      190.50    115.91  33
73      71    Weaver, Jason      T/G     77    305      195.58    138.64  25
74      60    Welch, Thomas        T     79    310      200.66    140.91  27
75      24     White, Corey       CB     73    205      185.42     93.18  24
```

	Experience	College
70	3	Tulsa
71	R	Jacksonville State
72	11	Georgia
73	1	Southern Mississippi
74	4	Vanderbilt
75	3	Samford

Observe that the covariances between height and weight are different, depending on the units of measurement as we discussed before, but the correlations are the same (within rounding error) for height and weight, regardless of the measurement units. The covariance is found with the cov() function, and the correlation with the cor() function.

```
> cov(saints$Inches, saints$Pounds)
[1] 80.86685
> cov(saints$centimeters, saints$kilograms)
[1] 93.36248
> cor(saints$Inches, saints$Pounds)
[1] 0.6817679
> cor(saints$centimeters, saints$kilograms)
[1] 0.6817578
```

For the height and weight data, there is a positive correlation. The cor.test() function tests the significance of the correlation coefficient:

```
> cor.test(saints$Inches, saints$Pounds)

        Pearson's product-moment correlation

data:  saints$Inches and saints$Pounds
t = 7.9624, df = 73, p-value = 1.657e-11
alternative hypothesis: true correlation is not equal to 0
95 percent confidence interval:
 0.5380635 0.7869594
sample estimates:
      cor
0.6817679
```

With 73 degrees of freedom, the correlation is highly significant. The astute reader will notice that the significance of the correlation is tested with a t test.

The correlation coefficient can be used to calculate the intercept and the slope of the regression line, as follows. We will symbolize the intercept as b_0 and the regression coefficient (slope) as b_1. The slope is

$$b_1 = r_{xy} \left(\frac{s_y}{s_x} \right).$$

and the intercept is

$$b_0 = \bar{y} - b_1 \bar{x}.$$

See that the intercept will be the mean of y when the slope coefficient is zero. The lm() function (for linear model) will produce the coefficients. The t test in the bivariate case for the slope coefficient is equivalent to the t test for the overall regression, and the F ratio for the regression is the square of the t value.

```
> model <- lm(saints$Pounds ~ saints$Inches)
> summary(model)

Call:
lm(formula = saints$Pounds ~ saints$Inches)

Residuals:
    Min      1Q  Median      3Q     Max
-68.877 -21.465  -1.282  19.351 104.579

Coefficients:
               Estimate Std. Error t value Pr(>|t|)
(Intercept)    -662.668    113.935  -5.816 1.48e-07 ***
saints$Inches    12.228      1.536   7.962 1.66e-11 ***
---
Signif. codes:  0 '***' 0.001 '**' 0.01 '*' 0.05 '.' 0.1 ' ' 1

Residual standard error: 33.97 on 73 degrees of freedom
Multiple R-squared:  0.4648,    Adjusted R-squared:  0.4575
F-statistic:  63.4 on 1 and 73 DF,  p-value: 1.657e-11
```

Recipe 11-2. Special Cases of the Correlation Coefficient
Problem

We often have data for x or y (or both) that are not interval or ratio, but we would still like to determine whether there is a relationship or association between the variables. There are alternatives to the Pearson correlation for such circumstances. We will examine the case in which one variable is measured on a scale while the other is measured as a dichotomy, where one or both variables are measured at the ordinal level, and where both variables are measured as dichotomies.

Solution

When you correlate a binary (0, 1) variable with a variable measured on a scale, you are using a special case of the correlation coefficient known as the *point-biserial correlation*. The computations are no different from those for the standard correlation coefficient. The binary variable might be the predictor x, or it might be the criterion y. When the binary variable is the predictor, the point-biserial correlation is equivalent to the independent-samples t test with regard to the information to be gleaned from the analysis. When the binary variable is the criterion, we are predicting a dichotomous outcome such as pass/fail. There are some issues with predicting a (0, 1) outcome using standard correlation and regression, and logistic regression offers a very attractive alternative that addresses these issues. When both variables are dichotomous, the index of relationship is known as the *phi coefficient*, and it has a relationship to chi-square, as you will learn.

Point-Biserial Correlation

Let us return to the example used for the independent-samples t test in Recipe 10-3. For this analysis, the data are reconfigured so that the dependent variable is the x scores followed by the y scores from the previous example. A 0 is used to indicate membership in x and a 1 is used to indicate y group membership.

```
> head(biserial)
  newDV indicator
1  60.2         0
2  64.4         0
3  64.8         0
4  65.6         0
5  63.7         0
6  68.0         0
> tail(biserial)
   newDV indicator
45  79.8         1
46  82.5         1
47  80.7         1
48  86.4         1
49  81.2         1
50  94.1         1
```

We can now regress the new dependent variable onto the column of zeros and ones by using the lm( ) function. The results will be instructive in terms of the relationship between the t test and correlation.

```
> biserialCorr <- lm(newDV ~ indicator, data = biserial)
> summary(biserialCorr)

Call:
lm(formula = newDV ~ indicator, data = biserial)

Residuals:
    Min      1Q  Median      3Q     Max
-16.616  -4.691  -0.120   5.059  19.876

Coefficients:
            Estimate Std. Error t value Pr(>|t|)
(Intercept)   65.824      1.647  39.958  < 2e-16 ***
indicator     18.692      2.330   8.024 2.03e-10 ***
---
Signif. codes:  0 '***' 0.001 '**' 0.01 '*' 0.05 '.' 0.1 ' ' 1

Residual standard error: 8.237 on 48 degrees of freedom
Multiple R-squared:  0.5729,    Adjusted R-squared:  0.564
F-statistic: 64.38 on 1 and 48 DF,  p-value: 2.029e-10
```

Now, compare the results from the linear model with the results of the standard t test assuming equal variances for x and y. Observe that the value of t testing the significance of the slope coefficient in the regression model is equivalent to the value of t comparing the means (ignoring the sign). The intercept term in the regression model is equal to the mean of x, and the slope coefficient is equal to the difference between the means of x and y. These equivalences are because both the correlation and the t test are based on the same underlying general linear model.

```
> t.test(x, y, var.equal = T)

        Two Sample t-test

data:  x and y
t = -8.0235, df = 48, p-value = 2.029e-10
alternative hypothesis: true difference in means is not equal to 0
95 percent confidence interval:
 -23.37607 -14.00793
sample estimates:
mean of x mean of y
   65.824    84.516
```

As you just saw, the use of a dichotomous predictor is equivalent to an independent-samples t test. However, when the criterion is dichotomous and the predictor is interval or ratio, problems often arise. The problem is that the regression equation can lead to predictions for the criterion that are less than 0 and greater than 1. These can simply be converted to 0s and 1s as an expedient, but a more elegant alternative is to use binary logistic regression rather than point-biserial correlation for this purpose. Let us explore these issues with the following data. The job satisfaction data we worked with in Chapter 10 have been modified to include two binary variables, which indicate whether the individual is satisfied overall (1 = yes, 0 = no) and whether the person was promoted (1 = yes, 0 = no).

```
> head(jobSat)
  subject satisf perf alternatives investment satYes promoYes
1       1     52   61          low         low      0        0
2       2     42   72          low         low      0        1
3       3     63   74          low         low      0        1
4       4     48   69          low         low      0        0
5       5     51   65          low         low      0        0
6       6     55   71          low         low      0        0
> tail(jobSat)
   subject satisf perf alternatives investment satYes promoYes
55      55     87   67         high        high      1        0
56      56     80   75         high        high      1        1
57      57     88   70         high        high      1        0
58      58     79   71         high        high      1        0
59      59     83   67         high        high      1        0
60      60     80   73         high        high      1        1
```

Say we are interested in determining the correlation between promotion and satisfaction. We want to know if those who are more satisfied are more likely to be promoted. Our criterion now is a dichotomous variable. We can regress this variable onto the vector of satisfaction scores using the lm() function. We see there is a significant linear relationship. Those who are more satisfied are also more likely to be promoted. The R^2 value indicates the percentage of variance overlap between satisfaction and promotion, which is approximately 14.8%. The product-moment correlation is the square root of the R^2 value, which is .385 because the slope is positive.

```
> predictPromo <- lm(promoYes ~ satisf, data = jobSat)
> summary(predictPromo)

Call:
lm(formula = promoYes ~ satisf, data = jobSat)

Residuals:
    Min      1Q  Median      3Q     Max
-0.6705 -0.4584 -0.1031  0.4240  0.8816

Coefficients:
             Estimate Std. Error t value Pr(>|t|)
(Intercept) -0.270122   0.229652  -1.176  0.24431
satisf       0.010225   0.003222   3.174  0.00241 **
---
Signif. codes:  0 '***' 0.001 '**' 0.01 '*' 0.05 '.' 0.1 ' ' 1

Residual standard error: 0.4652 on 58 degrees of freedom
Multiple R-squared:  0.148,     Adjusted R-squared:  0.1333
F-statistic: 10.07 on 1 and 58 DF,  p-value: 0.002409
```

Using the coefficients from our linear model, we can develop a prediction for each employee's promotion from his or her satisfaction score. The predict(lm) function performs this operation automatically.

```
> predict <- predict(lm(promoYes ~ satisf, data = jobSat))
> predict <- round(predict,2)
> jobSat <- cbind(jobSat, predict)
> head(jobSat)
  subject satisf perf alternatives investment satYes promoYes predict
1       1     52   61          low        low      0        0    0.26
2       2     42   72          low        low      0        1    0.16
3       3     63   74          low        low      0        1    0.37
4       4     48   69          low        low      0        0    0.22
5       5     51   65          low        low      0        0    0.25
6       6     55   71          low        low      0        0    0.29

> summary(jobSat$predict)
   Min. 1st Qu.  Median    Mean 3rd Qu.    Max.
 0.0200  0.2575  0.5100  0.4347  0.5800  0.6900
```

In the current case, the predicted values for promotion are all probabilities between 0 and 1. Note, however, that this is not always the case. In many situations, when you are predicting a binary outcome, the predicted values will be lower than 0 or greater than 1, as mentioned earlier. A preferable alternative in such cases is binary logistic regression. Use the glm() function (for general linear model) to perform the binary logistic regression. We create a model called myLogit and then obtain the summary.

```
> myLogit <- glm(promoYes ~ satisf, data = jobSat, family = "binomial")
> summary(myLogit)

Call:
glm(formula = promoYes ~ satisf, family = "binomial", data = jobSat)

Deviance Residuals:
    Min      1Q   Median      3Q     Max
-1.5272  -1.0785  -0.5136  1.0394  2.0143

Coefficients:
            Estimate Std. Error z value Pr(>|z|)
(Intercept) -3.77399    1.33523  -2.826  0.00471 **
satisf       0.04964    0.01791   2.771  0.00559 **
---
Signif. codes:  0 '***' 0.001 '**' 0.01 '*' 0.05 '.' 0.1 ' ' 1

(Dispersion parameter for binomial family taken to be 1)

    Null deviance: 82.108  on 59  degrees of freedom
Residual deviance: 72.391  on 58  degrees of freedom
AIC: 76.391

Number of Fisher Scoring iterations: 3
```

This model is also significant. We see that for approximately every unit increase in job satisfaction, the log odds of promotion versus nonpromotion increase by .05.

The Phi Coefficient

The phi coefficient is a version of the correlation coefficient in which both variables are dichotomous. We also use the phi coefficient as an effect-size index for chi-square for 2 × 2 contingency tables. In this case, correlating the columns of zeros and ones is an equivalent analysis to conducting a chi-square test on the 2 × 2 contingency table. The relationship between phi and chi-square is as follows, where N is the total number of observations. First, we calculate the standard correlation coefficient with the two dichotomous variables, and then we calculate chi-square and show the relationship.

$$\phi^2 = \frac{\chi^2}{N}.$$

```
> cor.test(jobSat$satYes, jobSat$promoYes)

        Pearson's product-moment correlation

data:  jobSat$satYes and jobSat$promoYes
t = 3.4586, df = 58, p-value = 0.001024
alternative hypothesis: true correlation is not equal to 0
95 percent confidence interval:
 0.1782873 0.6039997
sample estimates:
      cor
0.4134925
```

The value of chi-square (without the Yates continuity correction) is as follows:

```
> chisq.test(table(jobSat$satYes, jobSat$promoYes), correct = F)

        Pearson's Chi-squared test

data:  table(jobSat$satYes, jobSat$promoYes)
X-squared = 10.2586, df = 1, p-value = 0.001361
```

Apart from rounding error, the phi coefficient is equal to the correlation that we calculated previously.

```
> phi <- sqrt(10.2586/60)
> phi
[1] 0.4134932
```

The Spearman Rank Correlation

The Spearman Rank Correlation, r_s, (also known as Spearman's Rho) is the nonparametric version of the correlation coefficient for variables measured at the ordinal level. Technically, this index measures the degree of monotonic increasing or decreasing relationship instead of linear relationship. This is because ordinal measures do not provide information concerning the distances between ranks. Some data are naturally ordinal in nature, while other data are converted to ranks because the data are nonnormal in their distribution.

As with the phi coefficient, r_s can be calculated using the formula for the product moment coefficient after converting both variables to ranks. The value of r_s is calculated as follows:

$$r_s = 1 - \frac{6\sum d_i^2}{n(n^2 - 1)}.$$

where $d_i = x_i - y_1$ is the difference between ranks for a given pair of observations. The cor.test() function in R provides the option for calculating Spearman's r_s. To illustrate, let us examine the relationship between the job satisfaction and job performance. See that when we choose the argument method = "spearman" for the cor.test() function, the result is identical to the standard correlation between the ranks for the two variables.

```
> cor.test(jobSat$satisf, jobSat$perf, method = "spearman")

        Spearman's rank correlation rho

data:  jobSat$satisf and jobSat$perf
S = 20906.99, p-value = 0.0008597
alternative hypothesis: true rho is not equal to 0
sample estimates:
      rho
0.4190888

Warning message:
In cor.test.default(jobSat$satisf, jobSat$perf, method = "spearman") :
  Cannot compute exact p-value with ties
```

Now, examine the result of correlating the two sets of ranks:

```
> cor.test(rank(jobSat$satisf), rank(jobSat$perf))

        Pearson's product-moment correlation

data:  rank(jobSat$satisf) and rank(jobSat$perf)
t = 3.5153, df = 58, p-value = 0.0008597
alternative hypothesis: true correlation is not equal to 0
95 percent confidence interval:
 0.1848335 0.6082819
sample estimates:
      cor
0.4190888
```

Recipe 11-3. A Brief Introduction to Multiple Regression

Problem

In many research studies, there is a single dependent (criterion) variable but multiple predictors. Multiple correlation and regression apply in cases of one dependent variable and two or more independent (predictor or explanatory) variables. Multiple regression takes into account not only the relationships of each predictor to the criterion, but the intercorrelations of the predictors as well. The multiple linear regression model is as follows:

$$\hat{y} = b_0 + b_1 x_1 + \cdots + b_k x_k.$$

This means that the predicted value of y is a linear combination of an intercept term added to a weighted combination of the k predictors. The regression coefficients, symbolized by b, are calculated in such a way that the sum of the squared residuals (the differences between the observed and predicted values of y) is minimized. The multiple correlation coefficient, R, is the product-moment correlation coefficient between the observed and predicted y values. Just as in bivariate correlation and regression, R^2 shows the percentage of variation in the dependent (criterion) variable that can be explained by knowing the value of the predictors for each individual or object.

Finding the values of the regression coefficients and testing the significance of the overall regression allows us to determine the relative importance of each predictor to the overall regression, and whether a given variable contributes significantly to the prediction of y. We use multiple regression in many different applications.

Solution

Multiple regression is a very general technique. ANOVA can be seen as a special case of multiple regression, just as the independent-samples t test can be seen as a special case of bivariate regression. Let us consider a common situation. We would like to predict college students' freshman grade point average (GPA) from a combination of motivational and academic variables.

The following data are from a large-scale retention study I performed for a private liberal arts university. Data were gathered from incoming freshmen over a three-year period from 2005 to 2007. Academic variables included the students' high school GPAs, their freshman GPAs, their standardized test scores (ACT and SAT), and the students' rank in their high schools' graduating class, normalized by the size of the high school. Motivational variables included whether the student listed this university first among the schools to receive his or her standardized test scores (preference) and how many hours the student attempted in his or her first semester as a freshman. Another motivational variable was whether the student was the recipient of an athletic scholarship. Demographic variables included the students' sex and ethnicity. The data are as follows:

```
> head(gpa)
     id pref frYr frHrs female hsGPA collGPA retained ethnic SATV SATM  SAT ACT
1 17869    1 2006  15.5      1  4.04    2.45        1      1  480  530 1010  18
2 17417    1 2005  19.0      1  4.20    2.89        0      0  520  570 1090  20
3 16681    1 2006  15.0      1  4.15    2.60        1      1  400  410  810  18
4 17416    1 2005  16.5      1  4.50    3.39        1      1  490  480  970  23
5 16990    1 2005  17.0      1  3.83    2.65        1      1  440  460  900  18
6 16254    1 2006  15.0      1  4.32    1.93        1      1  450  650 1100  22
  hsRank hsSize classPos athlete
1    129    522     0.75       0
2     35    287     0.88       0
3      9     13     0.31       0
4      6    158     0.96       0
5     94    279     0.66       0
6     23    239     0.90       0
> tail(gpa)
       id pref frYr frHrs female hsGPA collGPA retained ethnic SATV SATM  SAT
290 17019    1 2006    17      1  3.76    1.57        1      1  480  590 1070
291 18668    1 2007    15      1  4.24    2.80        1      1  540  480 1020
292 18662    1 2007    15      0  3.31    2.33        1      1  440  570 1010
293 19217    1 2007    16      0  4.54    2.25        1      1  540  470 1010
294 16041    1 2005    16      0  3.55    2.69        1      1  540  570 1110
295 18343    1 2007    16      1  4.21    3.31        1      1  530  660 1190
    ACT hsRank hsSize classPos athlete
290  18    133    297     0.55       0
291  24     34    243     0.86       0
292  17    179    306     0.42       1
293  20      7     12     0.42       0
294  22    218    498     0.56       1
295  27     32    301     0.89       0
```

The lm( ) function in base R can be used for multiple regression. Let us create a linear model with the college GPA as the criterion and the academic variables of SAT verbal, SAT math, high school GPA, and class position as predictors. We can also add sex, preference, and status as a student athlete as predictors. We will run the full model and then eliminate any predictors that do not add significantly to the overall regression.

```
> mulReg <- lm(collGPA ~ SATV + SATM + hsGPA + classPos + female + pref + athlete, data = gpa)
> summary(mulReg)

Call:
lm(formula = collGPA ~ SATV + SATM + hsGPA + classPos + female +
    pref + athlete, data = gpa)

Residuals:
    Min      1Q  Median      3Q     Max
-1.2170 -0.3267  0.0130  0.3253  1.4529
```

```
Coefficients:
              Estimate Std. Error t value Pr(>|t|)
(Intercept)  0.2207168  0.2874792   0.768  0.44326
SATV         0.0015583  0.0005095   3.059  0.00243 **
SATM         0.0003460  0.0005027   0.688  0.49178
hsGPA        0.2620812  0.0986011   2.658  0.00830 **
classPos     1.0528486  0.2416664   4.357 1.84e-05 ***
female       0.1714455  0.0704189   2.435  0.01552 *
pref        -0.2986353  0.1164998  -2.563  0.01088 *
athlete      0.0574808  0.0772814   0.744  0.45761
---
Signif. codes:  0 '***' 0.001 '**' 0.01 '*' 0.05 '.' 0.1 ' ' 1

Residual standard error: 0.5136 on 287 degrees of freedom
Multiple R-squared:  0.4433,    Adjusted R-squared:  0.4297
F-statistic: 32.64 on 7 and 287 DF,  p-value: < 2.2e-16
```

The significant predictors are SAT verbal, high school GPA, class position, sex, and stated preference for this university. Let us run the regression again using only those predictors. Normally, we eliminate the nonsignificant variables one at a time; that is, we exclude the predictor with the highest p-value first, but here we eliminate both SATM and the athlete variable at the same time for illustrative purposes, as the p-values are quite similar.

```
> mulReg2 <- lm(collGPA ~ SATV + hsGPA + classPos + female + pref, data = gpa)
> summary(mulReg2)

Call:
lm(formula = collGPA ~ SATV + hsGPA + classPos + female + pref,
    data = gpa)

Residuals:
    Min     1Q  Median     3Q    Max
-1.2199 -0.3445  0.0213  0.3131  1.4816

Coefficients:
              Estimate Std. Error t value Pr(>|t|)
(Intercept)  0.3182779  0.2730307   1.166 0.244688
SATV         0.0016898  0.0004428   3.816 0.000166 ***
hsGPA        0.2663688  0.0960505   2.773 0.005912 **
classPos     1.0850437  0.2395356   4.530 8.65e-06 ***
female       0.1517149  0.0680608   2.229 0.026575 *
pref        -0.3008454  0.1152196  -2.611 0.009497 **
---
Signif. codes:  0 '***' 0.001 '**' 0.01 '*' 0.05 '.' 0.1 ' ' 1

Residual standard error: 0.5129 on 289 degrees of freedom
Multiple R-squared:  0.4408,    Adjusted R-squared:  0.4312
F-statistic: 45.57 on 5 and 289 DF,  p-value: < 2.2e-16
```

The value of R^2 is basically unchanged, and our new model uses only predictors that each contribute significantly to the overall regression. The predictors are highly interrelated, as the following correlations show, so we are interested in finding a combination of predictors that are each related to the criterion independently.

We can apply the regression equation we created to get predicted values of the college GPA, along with a 95% confidence interval using the `predict.lm` function. Remember the product-moment correlation between the predicted *y* values and the observed *y* values is multiple *R*, as the following output shows. The `predict.lm` function will optionally provide a confidence interval for each fitted value of *y*, giving the lower and upper bounds for the mean *y* for a given *x*. You can also specify a prediction interval, which will give the prediction limits for the range of individual *y* scores for a given *x* score. See that the `predict.lm()` function produces a matrix. Squaring the correlation between the fitted values and the observed *y* values produces the value of R^2 reported earlier.

```
> gpa.fit <- predict.lm(mulReg2, interval = "confidence")
> head(gpa.fit)
       fit      lwr      upr
1 2.870169 2.788140 2.952199
2 3.121436 3.038125 3.204747
3 2.286866 2.010149 2.563582
4 3.237456 3.128370 3.346542
5 2.648985 2.553701 2.744270
6 3.056815 2.945514 3.168115
> cor(gpa.fit[,1], gpa$collGPA)^2
[1] 0.4408457
```

Our linear model is as follows.

$$\hat{y} = 0.3183 + .00169(\text{SATV}) + .26637(\text{hsGPA}) + 1.085(\text{classPos}) + .1517(\text{female}) - 0.300845(\text{pref}).$$

This formula produces the fitted value of 2.8702 as the predicted value of the college GPA for the first student in the dataset. The interpretation of the confidence interval is that we are 95% confident that students with the same SAT verbal score, the same high school GPA, the same class position, and the same stated preference as this student will have college GPAs between 2.788 and 2.952.

CHAPTER 12

■ ■ ■

Contemporary Statistical Methods

You learned the traditional tests for differences in Chapter 10 and for relationships in Chapter 11. In Chapter 12, you will learn some of the modern statistical methods developed over the past half century. You have learned by now that statistics is not a static field, but a growing, dynamic, and evolving one. With modern technology, statisticians are able to perform calculations and analyses not previously available. The notion of sampling from a population and estimating parameters is still an important one, but newer techniques that allow us to use samples in new ways are increasingly available.

In Chapter 12, you will learn about modern robust statistical techniques that allow us to resample our data repeatedly and build the distribution of a statistic of interest, perhaps even one of our own creation. We can also simulate data and make inferences concerning various statistics. We can use techniques that improve our ability to make correct inferences when data distributions depart from normality. It is also possible to use permutation tests as alternatives to traditional hypothesis tests.

The disenchantment with null hypothesis significance testing began soon after R. A. Fisher suggested the technique in the first place. Fisher asserted that one does not (ever) need an alternative hypothesis. Instead, one examines a sample of data under the assumption of some population distribution, and determines the probability that the sample came from such a population. If the probability is low, then one rejects the null hypothesis. If the probability is high, the null hypothesis is not rejected. The notion of an alpha or significance level was important to Fisher's formulation, but the idea of a Type II error was not considered. Fisher also developed the notion of a p value as an informal, but useful, way of determining just how likely it was that the sample results occurred by chance alone.

Karl Pearson's son Egon Pearson and Jerzy Neyman developed a different form of hypothesis testing. In their framework, two probabilities were calculated, each associated with a competing hypothesis. The hypothesis that was chosen had the higher probability of having produced the sample results. Thus, the Neyman-Pearson formulation considered (and calculated) both Type I and Type II errors. Fisher objected to the Neyman-Pearson approach, and a bitter rivalry ensued. The rivalry ended with Fisher's death in 1962, but the controversy endures.

Current hypothesis testing is an uneasy admixture of the Fisher and Neyman-Pearson approaches. The debate between these two approaches has been pursued on both philosophical and mathematical grounds. Mathematicians claim to have resolved the debate, while philosophers continue to examine the two approaches independently. However, an alternative is emerging that may soon supplant both of these approaches. These newer methods provide one the ability to make better, that is, more accurate and more powerful, inferences when data do not meet parametric distributional assumptions such as normality and equality of variance.

Recipe 12-1. Resampling Techniques
Problem

Many statistical problems are intractable from the perspective of traditional hypothesis testing. For example, making inferences about the distribution of certain statistics is difficult or impossible using traditional methods. Examine the histogram in Figure 12-1, which represents the hypothetical scores of 20 elderly people on a memory test. The data are clearly not normal. As the histogram indicates, some of the subjects have lost cognitive functioning. The problem is how to develop a confidence interval for the *median* of the data. It is clear how to develop a confidence interval for a mean, but what about a confidence interval for the median? Here are the 20 memory scores.

```
Memory <- c(2.5, 3.5, 4.5, 5.5, 6.5, 6.5, 6.5, 8.5, 8.5, 9.5, 9.5, 9.5, 9.5, 10.5, 10.5, 10.5, +
10.5, 11.5, 11.5, 12.5)
> breaks <- seq(2, 14, 1)
> hist(memory, breaks)
```

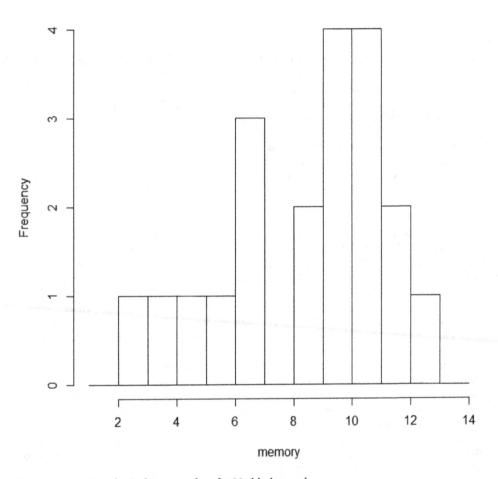

Figure 12-1. *Hypothetical memory data for 20 elderly people*

Solution

The solution in not as difficult as one might imagine. We treat the sample of 20 as a "pseudo-population," and take repeated resamples with replacement from our original sample. We then calculate the median of each sample and study the distribution of sample medians. We can find the 95% confidence limits from that distribution using R's quantile() function. Let us generate 10,000 resamples and calculate the median of each. Then we will generate a histogram of the distribution of the medians, followed by the calculation of the limits of the 95% confidence interval for the median. There are many ways to do this, but we can simply write a few lines of R code that loop explicitly through 10,000 iterations, taking a sample with replacement each time, and calculating and storing the median of each sample. Here is our R code. The numeric() command makes an empty vector that we then fill with the medians from each of the 10,000 resamples. The sampling with replacement is accomplished by setting the replace argument to T or TRUE.

```
nSamples <- 10000
medianSamp <- numeric()
for (i in 1:nSamples) medianSamp[i] <- median(sample(memory, replace = T))
hist(medianSamp)
(llMedian <- quantile(medianSamp, 0.025))
(ulMedian <- quantile(medianSamp, 0.975))
```

The histogram of the medians appears in Figure 12-2. Just as the original sample was negatively skewed, so is our distribution of medians.

Histogram of medianSamp

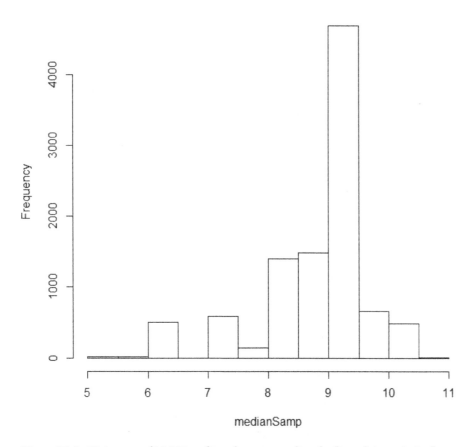

Figure 12-2. Histogram of 10,000 medians from resampling the "pseudo-population"

The 95% confidence interval for the median is [6.5, 10.5].

```
> (llMedian <- quantile(medianSamp, 0.025))
2.5%
 6.5
> (ulMedian <- quantile(medianSamp, 0.975))
 97.5%
10.5
```

The essence of all resampling techniques is the same as what we have done here. We use a sample of data and take a large number of resamples with replacement. We then study the distribution of the statistic of interest. This general approach is known as *bootstrapping*. The bootstrapping method can be used for other statistics simply by altering the R code. For example, we can study the distribution of the sample variance, the mean, or any other statistic we want without assuming anything about the population.

Recipe 12-2. Making Inferences About Means from Real Data
Problem
Real data are very unlikely to be normally distributed. Indeed, the parametric assumptions for *t* tests and ANOVA are often not met. When we make inferences about means and are unwilling to assume the population is normal, or that the variances in the population are equal, modern robust statistical methods can be very helpful. This approach was championed and described by Rand Wilcox in his articles and books on contemporary statistical methods. Wilcox has developed the WRS (Wilcox' Robust Statistics) package, with more than 1,100 functions for various robust statistical calculations. The WRS package is not available through CRAN because of the lack of complete documentation for all functions, but it can be obtained from GitHub or through Wilcox's web site at http://dornsife.usc.edu/labs/rwilcox/software/.

You will need to download this package if you wish to try out some of the functions that are used in the examples that follow. In general, a statistic is said to be *robust* to the extent that it performs well when the data are not normally distributed or when two or more samples have widely differing variances. Another concern with real data is that statistics such as the mean are heavily influenced by the presence of outliers. The median, on the other hand, is a robust statistic for estimating the central tendency of a data distribution because it is unaffected by extreme values. To elaborate further on the idea of robustness, Mosteller and Tukey defined two types, *resistance* and *efficiency*. A statistic (for example, the median) is resistant if changing a small part of the data (even by a large amount) does not lead to a large change in the estimate. A statistic has robustness of efficiency, on the other hand, if the statistic has high efficiency in a variety of situations, rather than a single situation. For example, the mean and variance, which are used as the basis for standard confidence intervals, do not have robustness of efficiency because they work well in this capacity only when data are normally distributed or when sample sizes are large.

Solution

The standard *t* test discussed in Chapter 10 is useful for comparing means when the data are sampled from normal distributions and when the variances are equal in the population. The Welch *t* test (the R default) is a better alternative when the population variances cannot be assumed to be equal. However, both versions of the *t* test are ineffective when the data differ substantially from a normal distribution—that is, when the distributions are highly skewed, as is often the case with real data (especially if the data have not been cleaned).

Trimmed means remove outliers by excluding a certain percentage of values at the high and low ends of the data distribution, and then calculating the mean of the remaining data values. Similarly, we can calculate Winsorized statistics. Winsorizing (named after Charles P. Winsor, who suggested the technique) is slightly more complicated than trimming. Winsorizing is defined as follows: the *k*-th Winsorized mean is the average of the observations after each of the first k smallest values are replaced by the $(k + 1)$th smallest value, and the k largest values are replaced by the $(k + 1)$th largest value. Thus, Winsorizing does not discard data values, as trimming does, but replaces data values with other observed values that are closer to the center of the data distribution.

The mean and the variance are used as the basis of confidence intervals, and by our definitions, they are efficient when the data are from a normal population. However, these statistics do not possess robustness of efficiency or resistance. Therefore, we would like to find estimators of the population mean that are both resistant and possess robustness of efficiency. We can develop confidence limits for a trimmed mean or for the differences between trimmed means. These techniques are robust alternatives to the standard one-sample and two-sample t tests.

Calculating and examining trimmed means is one alternative to the standard or Welch t test. A common approach is to use a 10% trimmed mean, therefore excluding the upper 10% and the lower 10% of observations. The mean function in R can be used to calculate the trimmed mean of a set of numbers. For example, let us calculate the trimmed mean for the numbers we used in Recipe 12-1. With 20% trimming, the trimmed mean is closer to the median than the original mean was, though even with trimming, the trimmed data are still negatively skewed, as the trimmed mean is lower than the median. The median is resistant, but has less robustness of efficiency than a trimmed or Winsorized mean because these statistics use more information than does the median.

We can calculate the variance for a trimmed mean as follows:

$$\frac{s_w^2}{n(1-2\gamma)^2}$$

where σ_w^2 is the Winsorized variance as discussed earlier, n is the sample size before trimming, and γ is the proportion of trimming. We can then calculate the standard error of the trimmed mean as

$$\frac{s_w}{\sqrt{n}(1-2\gamma)}$$

Wilcox's WRS package includes a tmean function for calculating the trimmed mean, a trmse function for estimating the standard error of the trimmed mean, a win function for calculating a Winsorized mean, and a winvar function for calculating a Winsorized variance. The following code shows these functions applied to the memory data from Recipe 12-1. Observe that the built-in mean function produces the same answer as Wilcox's tmean function, where the trimming proportion is the same. Note also that the trimming substantially reduced the variance of the trimmed data.

```
> mean(memory)
[1] 8.4
> mean(memory, trim = .2)
[1] 8.833333
> tmean(memory, tr = .2)
[1] 8.833333
> trimse(memory, tr = .2)
[1] 0.6578363
> win(memory, tr = .2)
[1] 8.7
> winvar(memory, tr = .2)
[1] 3.115789
> var(memory)
[1] 7.989474
```

We can also get a confidence interval for the trimmed mean by using Wilcox's function `trimci`. This can substitute for a one-sample t test when the data are not normally distributed. If the test value (the hypothesized population mean) is within the limits of the confidence interval, we conclude that the evidence suggests that the sample came from a population with μ_0.

```
> trimci(memory, tr = .2, alpha = .05)
[1] "The p-value returned by the this function is based on the"
[1] "null value specified by the argument null.value, which defaults to 0"
$ci
[1]   7.385445 10.281221

$estimate
[1] 8.833333

$test.stat
[1] 13.42786

$se
[1] 0.6578363

$p.value
[1] 3.63381e-08

$n
[1] 20
```

Now, assume that we locate a sample of elderly people who take a memory-enhancing supplement. We want to compare the means of the two groups (labeled memory1 and memory2). We note that the memory scores for the first group are negatively skewed and that the scores for the second group are positively skewed (see Figure 12-3). We find further that the variances of the two groups are not equal.

```
> memory1
 [1]   2.5  3.5  4.5  5.5  6.5  6.5  6.5  8.5  8.5  9.5  9.5  9.5  9.5 10.5 10.5
[16] 10.5 10.5 11.5 11.5 12.5
> memory2
 [1]   5.5  5.5  6.0  6.0  6.0  7.5  7.5  7.5  7.5  7.5  7.5 10.0 10.0 11.0 13.0
[16] 14.0 16.0 16.0 16.0 20.0
>
```

```
[1] 8.4
> mean(memory2)
[1] 10
> var(memory1)
[1] 7.989474
> var(memory2)
[1]  18.94737
```

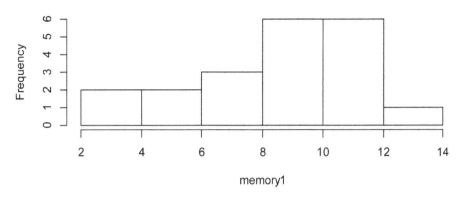

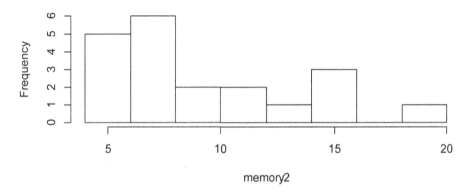

Figure 12-3. *Comparison of memory scores of two groups of elderly persons*

A robust alternative to the two-sample *t* test was developed by Yuen. The Yuen *t* statistic is the difference between the two trimmed means divided by the estimated standard error of the difference between the trimmed means,

$$t_y = \frac{\overline{X}_{t1} - \overline{X}_{t2}}{\sqrt{d_1 + d_2}}$$

where the standard error estimates,

$$d_j = \frac{(n_j - 1)\sigma_{wj}^2}{h_j(h_j - 1)},$$

and the variance estimate, σ_{wj}^2, is the γ-Winsorized variance for group j, and h_j is the effective sample size after trimming. The degrees of freedom for the Yuen test are calculated as

$$\upsilon_y = \frac{(d_1 + d_2)^2}{d_1^2 / (h_1 - 1) + d_2^2 (h_2 - 1)}.$$

Wilcox's function yuen performs this test. Let us compare the results of the Welch test and the Yuen test to determine the effect of trimming on the resulting confidence intervals.

```
> t.test(memory1, memory2)

        Welch Two Sample t-test

data:  memory1 and memory2
t = -1.3787, df = 32.604, p-value = 0.1774
alternative hypothesis: true difference in means is not equal to 0
95 percent confidence interval:
 -3.9622155  0.7622155
sample estimates:
mean of x mean of y
      8.4      10.0
```

The yuen function produces a list as its output, so we will create an object and query the results for the confidence interval and *p*-value. The yuen function defaults to a trimming value of 0.20 and an alpha level of 0.05.

```
> yuenTest <- yuen(memory1, memory2)
> yuenTest$ci

[1] -3.215389  2.715389
> yuenTest$p.value
[1] 0.8609031
```

When the sample size is small, as it is in this case, a preferable alternative is to use a bootstrapped version of Yuen's robust *t* test. Wilcox provides a function yuenbt to calculate a percentile-*t* bootstrap that estimates a confidence interval for the difference between two trimmed means. As with any hypothesis testing situation where the hypothesized difference is zero, the fact that zero or no difference is "in" the confidence interval implies that we do not reject the null hypothesis.

```
> yuenbt(memory1, memory2, nboot = 2000)
[1] "NOTE: p-value computed only when side=T"
$ci
[1] -3.940663  2.441760

$test.stat
[1] -0.1811149

$p.value
[1] NA

$est.1
[1] 8.833333

$est.2
[1] 9.083333
```

```
$est.dif
[1] -0.25

$n1
[1] 20

$n2
[1] 20
```

The robust tests and confidence intervals indicate the two groups are more similar than does the standard parametric approach. As a last alternative, we can simply use a nonparametric test, the Mann-Whitney U test, to determine if the two distributions differ in shape. This test also shows that the memory scores of the two groups do not differ significantly.

```
> wilcox.test(memory1, memory2)

        Wilcoxon rank sum test with continuity correction

data:  memory1 and memory2
W = 177, p-value = 0.5417
alternative hypothesis: true location shift is not equal to 0

Warning message:
In wilcox.test.default(memory1, memory2) :
  cannot compute exact p-value with ties
```

Recipe 12-3. Permutation Tests
Problem

Permutation tests are similar in logic to bootstrap tests, but use permutations of the data rather than resampling with replacement. Because of this, a permutation test repeated on the same data will produce the same results, whereas a bootstrap test with the same data will produce different results. In many cases, the two will be similar, but remember because bootstraps use sampling with replacement, no two bootstrap tests will be entirely equivalent. On the other hand, techniques that use permutation tests will always produce the same results because of the nature of permutations. When an exact answer is needed, permutation tests are preferred over bootstrap tests.

The permutation test was developed by R. A. Fisher in the 1930s. At that time, permutations had to be calculated by hand, a tedious process indeed. Fisher's famous "lady and the tea" problem illustrates the logic of permutation tests. Fisher met a lady (Dr. Muriel Bristol) who claimed to have such a discerning sense of taste that she should correctly distinguish between tea into which milk had been poured before the tea and tea into which the milk was poured into the cup after the tea. Fisher wanted to test the hypothesis that this woman was not able to guess correctly more than at a chance level (the null hypothesis). The test was performed by pouring four cups each of tea with the milk poured first and with the tea poured first. The order of the eight cups was randomized, and each was presented to the lady for her judgment. The lady was then to divide the cups into two groups by the "treatment"—that is, whether milk or tea was added first. There are $_8C_4 = 70$ ways in which four cups could be chosen from eight cups. It has been reported that the lady correctly identified all 8 cups, but in his description of this experiment in his book, Fisher illustrated the test with three correct guesses and one incorrect guess. The critical region Fisher used for his test was $1/70 = .014$, which would occur only when the lady correctly identified all eight cups. The probability of identifying six or more of the eight cups was too high to justify rejecting the null hypothesis.

Solution

All permutation tests share the logic of Fisher's exact test. The perm package provides permutation tests for 2 and *K* samples. For example, we can perform a permutation test comparing the means of our two memory groups as follows. Either an asymptotic or an exact test can be conducted, though in our current case, the two produce virtually identical results.

```
> permTS(memory1, memory2)

        Permutation Test using Asymptotic Approximation

data:  memory1 and memory2
Z = -1.3505, p-value = 0.1769
alternative hypothesis: true mean memory1 - mean memory2 is not equal to 0
sample estimates:
mean memory1 - mean memory2
                       -1.6

> permTS(memory1, memory2, exact = T)

        Exact Permutation Test Estimated by Monte Carlo

data:  memory1 and memory2
p-value = 0.18
alternative hypothesis: true mean memory1 - mean memory2 is not equal to 0
sample estimates:
mean memory1 - mean memory2
                       -1.6
```

Writing Reusable Functions

R provides users the ability to write, save, and reuse functions. These functions are flexible and extend the power of R for users who need specialized capabilities not provided by the functions in base R or in contributed packages. Users are advised, however, to check CRAN (the Comprehensive R Archive Network) before they write a function, as it is very likely that someone else has encountered the same problem previously, and has already written the desired function for you. A case in point is the z test function we wrote in Recipe 10-1. As it materializes, the BSDA package already has a function labeled z.test that accomplishes the purpose we needed, and which also provides more options, including two-sample tests and directional and nondirectional alternative hypotheses.

Nonetheless, there usually comes a time when an R user wants to learn to be an R programmer, and writing functions is a key part of R programming. Functions do not have to be elegant or sophisticated to be useful. An often-used acronym in computer science is DRY (*don't repeat yourself*). Whenever you can reuse a line or lines of code to do something you need done more than once, functions will save you time and make your work more consistent.

Recipe 13-1. Understanding R Functions
Problem

Most users of R start as novices and then advance to users, programmers, and ultimately for some, contributors to the R community. Functions are a basic building block in R, and even if you are not particularly interested in writing functions, you still need to understand how functions work in order to become a more effective user of R, an R programmer, and perhaps ultimately a contributor of your own functions to CRAN.

As with everything else in R, a function is an object. Functions take input, perform operations on that input, and produce output. Functions can also call other functions and even take functions as input and produce different functions as output. You can create functions that ask for user input in the interactive R console, too.

Solution

Every R function has three components. These are the function *body* (the code inside the function), the function *arguments*, and the function *environment* (the locations of the function's variables). The function's arguments are both formal and actual. For example, a function to calculate the square of x has x as a formal argument, and may have the value of 8 as an actual argument:

```
> squareX <- function (x) {
+ print(x^2)
+ }
> squareX(8)
[1] 64
```

R evaluates the arguments to a function in a specified order. First, it looks for exact matches to the argument name (also called a tag). Second, R looks for any partial matches on tags. The formal arguments may also contain '…' as an indication that there are unspecified arguments. If the formal arguments contain the ellipsis '…', then any argument after it must have exact matching. The third pass is positional matching. Any unmatched formal arguments are "bound" to unnamed supplied arguments in order.

Functions arguments in R are lazy, which means that they are not evaluated if they are not used, and R will not produce an error if an unneeded argument is supplied, though it will often provide a warning. Let us use a common function to see how arguments are evaluated. As implied earlier, the order is unimportant if you use the argument names and exact or partial matches. Here is the mad function in R (for median absolute deviation from the median of a dataset). See that when you enter the function name without parentheses or arguments, R will print the function source code or other information, along with the function's bytecode location and environment namespace. I chose the mad function because it is one in which the code is written in R. Many other functions, such as sum, are written in C or some other language, and typing in the function name is much less instructive. See the following code for an illustration.

```
> mad
function (x, center = median(x), constant = 1.4826, na.rm = FALSE,
    low = FALSE, high = FALSE)
{
    if (na.rm)
        x <- x[!is.na(x)]
    n <- length(x)
    constant * if ((low || high) && n%%2 == 0) {
        if (low && high)
            stop("'low' and 'high' cannot be both TRUE")
        n2 <- n%/%2 + as.integer(high)
        sort(abs(x - center), partial = n2)[n2]
    }
    else median(abs(x - center))
}
<bytecode: 0x000000000c353650>
<environment: namespace:stats>


> sum
function (..., na.rm = FALSE)  .Primitive("sum")
```

Returning to the mad function, note the variety of arguments with defaults. The default value of "center" is the median of the dataset. There is a scaling constant, 1.4826, which ensures consistency with the principle that when x is distributed as $N(\mu, \sigma^2)$ with a large sample size, the expected value of the mean absolute deviation should approach the population standard deviation, σ. The na.rm argument tells R what to do if the dataset has missing values. The high and low arguments specify whether the "high median" or the "low median" will be used when the dataset has an even number of values. The high median is the higher of the two middle values, and the low median is the lower of the two middle values. With both of these arguments set to FALSE, the standard definition of the median as the average of the two middle values is used.

The following data are the serum cholesterol levels of 62 of the patients in the famous Framingham heart study:

```
> chol
 [1] 393 353 334 336 327 308 300 300 283 285 270 270 272 278 278 263 264 267 267
[20] 267 268 254 254 254 256 256 258 240 243 246 247 248 230 230 230 230 231 232
[39] 232 232 234 234 236 236 238 220 225 225 226 210 211 212 215 216 217 218 202
[58] 202 192 198 184 167
```

Let's use the mad function and see what happens when we omit all the arguments except x, the numeric vector.

```
> mad(chol)
[1] 35.5824
```

Now, we can specify the center as mean. Here are several different ways we could specify the argument, as we discussed. We can use an exact match to the argument name, omit it entirely because center is the second argument, or use a shorter version of the name for a partial match. All three approaches produce the same result.

```
> mad(chol, center = mean(chol))
[1] 29.74765
> mad(chol, mean(chol))
[1] 29.74765
> mad(chol, cen = mean(chol))
[1] 29.74765
```

As an extended example of writing a function, let us revisit the z.test function we worked with in Chapter 10. We will give our function the descriptive name of oneSampZ to distinguish it from the more general z.test function in the BSDA package. In the process of updating our function, we will modify it to make it more flexible, and you will learn how such improvements might work for you with similar kinds of problems.

First, let us clean up our function a bit by changing some of the labels and making the output a little nicer to look at. Following is the improved function for a one-sample z test. Remember we are doing a two-tailed test and developing a $(1 - \alpha) \times 100\%$ confidence interval. As you examine the function, identify the function body and the arguments. See that you can specify a default value for any of the arguments you like, as we did here for the alpha level. If you accept the default value, you can omit that particular argument when you call the function. A note is in order about the test value, which we call *mu* in keeping with the built-in t.test function. The actual hypotheses being tested are better described as follows for the function in its current state (we will expand on that momentarily). The usual notation for the null hypothesis is H_0, and for the alternative hypothesis, H_1. We are testing a null hypothesis that the population mean is equal to a known or hypothesized population mean, which we can label *m0*. Symbolically, our competing hypotheses are as follows:

$$H_o : \mu = \mu_0$$
$$H_1 : \mu \neq \mu_0$$

```
oneSampZ <- function(x, mu, sigma, n, alpha = 0.05){
        sampleMean <- mean(x)
        stdError <- sigma/sqrt(n)
        zcrit <- qnorm(1 - alpha/2)
        zobs <- (sampleMean - mu)/stdError
        LL <- sampleMean - zcrit * stdError
        UL <- sampleMean + zcrit * stdError
        pvalue <- 2 * (1 - pnorm(abs(zobs)))
        cat("\t","one-sample z test","\n","\n")
        cat("sample mean:",sampleMean,"\t","sigma:",sigma,"\n")
        cat("sample size:",n,"\n")
        cat("test value:",mu,"\n")
        cat("sample z:",zobs,"\n")
        cat("p value:",pvalue,"\n")
        cat((1-alpha),"percent confidence interval:","\n")
        cat("lower:",LL,"\t","upper:",UL,"\n")
        }
```

In the current case, with a two-sided hypothesis test, the confidence interval will be symmetrical around the mean, with half of the alpha in each tail of the normal distribution. To explain in detail how the function works, note that we pass a vector of scores, x, the test value mu sigma, the sample size, and the alpha level, which defaults to .05, to the function. The function calculates the sample mean, the standard error of the mean, the critical value of z, the sample z, the lower and upper limits of the confidence interval, and the two-tailed p value. These results are reported by use of the cat() function.

Let's add some features to our function. We can eliminate the need to pass the length of the vector by using the length() function for that purpose. We can also provide the option for the test to be one-tailed as well as two-tailed. The option to make the test one-tailed means we must allow for both right-tailed and left-tailed alternative hypotheses. As a result, the one-sided confidence intervals will not be symmetrical around the mean, but will have -∞ as the lower bound for the left-tailed test and +∞ as the upper bound for the right-tailed test. We need to add logic to our function to accommodate these choices and their consequences, so the function will necessarily become somewhat more complicated than before.

```
###########################################################################
#                                                                         #
#                     One-sample z Test Function                          #
#                        Written by Larry Pace                            #
#                                                                         #
###########################################################################
oneSampZ <- function(x, mu, sigma = sd(x), alternative = "two.sided", alpha = .05){
  meanX <- mean(x)
  n <- length(x)
  stdErr <- sigma/sqrt(n)
  zobs <- (meanX - mu)/stdErr
  choices <- c("two.sided","less","greater")
  chosen <- pmatch(alternative, choices)
  alternative <- choices[chosen]
  if (alternative == "greater") {
       testtype <- "alternative: true mean > test value"
       zcrit <- abs(qnorm(alpha))
       LL <- meanX - zcrit*stdErr
       UL <- Inf
       pvalue <- 1-pnorm(abs(zobs))
       }
  else if (alternative == "less") {
       testtype <- "alternative: true mean < test value"
       zcrit <- qnorm(1 - alpha)
       LL <- -Inf
       UL <- meanX + zcrit*stdErr
       pvalue <- pnorm(abs(zobs))
       }
  else {
       testtype <- "alternative: true mean unequal to test value"
       zcrit <- qnorm(1-alpha/2)
       LL <- meanX - zcrit*stdErr
       UL <- meanX + zcrit*stdErr
       pvalue <- 2*(1-pnorm(abs(zobs)))
       }
  cat("\t","one-sample z test","\n","\n")
  cat("sample mean:",meanX,"\n")
  cat("test value:",mu,"\n")
```

```
    cat("sigma:",sigma,"\n")
    cat(testtype,"\n")
    cat("observed z:",zobs,"\n")
    cat("p value:",pvalue,"\n")
    cat((1-alpha)*100," percent confidence interval","\n")
    cat("lower:",LL,"\t","upper:",UL,"\n")
}
```

This improved function takes a vector of scores, a test value, and a population standard deviation. If the population standard deviation is not provided, the function will use the sample standard deviation instead. The function defaults to a two-tailed test with an alpha level of .05. You can specify a right-tailed ("greater") or a left-tailed ("less") alternative hypothesis, and can change the alpha level if you choose. The function determines the choice of test and returns the appropriate p value along with a $(1-\alpha) \times 100\%$ confidence interval.

The hypotheses for the left-tailed and right-tailed tests are such that the confidence intervals are no longer symmetrical. There are two points of view about the null hypothesis in this case. Some statisticians continue to state the null hypothesis as a point estimate of the population parameter, while others modify the statement of the null hypothesis to encompass the entire range of possible outcomes. I tend to side with the first group, but many of my colleagues do not. Here is how we would state the hypotheses in both of these cases for a right-tailed test. First, the point-estimate statement:

$$H_0 : \mu = \mu_0$$
$$H_1 : \mu > \mu_0$$

Now the "cover all outcomes" approach, which emphasizes that the null hypothesis must cover all possibilities not represented by the alternative hypothesis:

$$H_0 : \mu \leq \mu_0$$
$$H_1 : \mu > \mu_0$$

We can use the fBasics package to get summary statistics for the cholesterol data we used for the z test.

```
> library(fBasics)
Loading required package: MASS
Loading required package: timeDate
Loading required package: timeSeries

Attaching package: 'fBasics'

The following object is masked from 'package:base':

    norm

> basicStats(chol)
                    chol
nobs          62.000000
NAs            0.000000
Minimum      167.000000
Maximum      393.000000
1. Quartile  225.250000
3. Quartile  267.750000
Mean         250.064516
Median       241.500000
```

Sum	15504.000000
SE Mean	5.258363
LCL Mean	239.549769
UCL Mean	260.579263
Variance	1714.323638
Stdev	41.404392
Skewness	1.002676
Kurtosis	1.439256

We note the data are positively skewed and positively kurtotic as the summary statistics show, and as we can visually verify in the histogram shown in Figure 13-1.

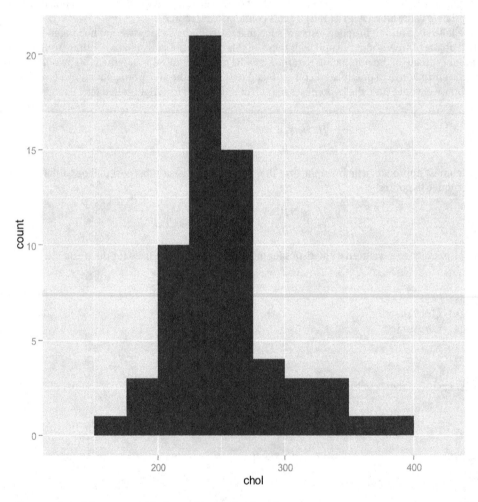

Figure 13-1. *Histogram of cholesterol levels for 62 patients*

With a sample size of 62, we should be able to invoke the central limit theorem and assume that the sampling distribution of means will be roughly normal. This would justify the use of a *z* test to compare this sample to the known mean and standard deviation of serum cholesterol levels in 20- to 74-year-old males, which are 211 and 46, respectively. To double-check the reasonability of our assumptions, we can perform a bootstrap sampling of our data to determine if the distribution of means is indeed approximately normal. See the plot of the sample means in Figure 13-2. For this plot, I converted the means to standard scores using the scale function, and then plotted the *z* scores with the normal curve as an overlay. Examine the following code to see how to do this. The freq = FALSE argument specifies the plotting of densities rather than frequencies, and the ylim argument makes sure there is enough room on the graph for the normal curve.

```
nsamps <- 10000
cholMean <- numeric()
for(i in 1:nsamps) cholMean[i] <- mean(sample(chol, replace = T))
CI <- c(quantile(cholMean, 0.025), quantile(cholMean, 0.975))
CI
    2.5%    97.5%
240.3867 260.6133

> zcholMean <- scale(cholMean)
> hist(zcholMean, freq = FALSE, col = "gray", ylim = c(0, 0.4))
> curve(dnorm, col = "red", add = TRUE)
```

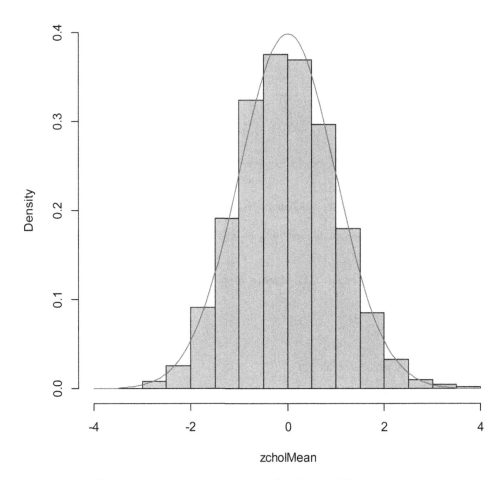

Histogram of zcholMean

Figure 13-2. *Plot of 10,000 resampled means from the cholesterol data*

Let us use our modified one-sample *z* test function to test the null hypothesis that the sample came from a population with a mean serum cholesterol level of 211 and a standard deviation of 46 against the one-sided alternative that the sample came from a population with a higher mean cholesterol level. We will adopt an alpha level of .01 for the test.

Here is the output from the one-sample *z* test, followed by the output from R's t.test function for comparison. The results are similar because the sample is relatively large.

```
> oneSampZ(chol, mu = 211, sigma = 46, alternative = "greater", alpha = 0.01)

        one-sample z test

sample mean: 250.0645
test value: 211
sigma: 46
alternative: true mean > test value
observed z: 6.686833
p value: 1.140266e-11
99  percent confidence interval
lower: 236.474    upper: Inf


> t.test(chol,mu = 211, alternative = "greater", conf.level = 0.99)

        One Sample t-test

data:  chol
t = 7.429, df = 61, p-value = 2.117e-10
alternative hypothesis: true mean is greater than 211
99 percent confidence interval:
 237.502     Inf
sample estimates:
mean of x
 250.0645
```

The extremely low *p* values indicate we can reject the null hypothesis, and we conclude that the sample came from a population with a mean higher than 211.

For comparison's sake, I also conducted the one-sample *z* test using the built-in z.test function in the BSDA package:

```
> library(BSDA)
> z.test(x = chol, alternative = "greater", mu = 211, sigma.x = 46, conf.level = 0.99)

        One-sample z-Test

data:  chol
z = 6.6868, p-value = 1.14e-11
alternative hypothesis: true mean is greater than 211
99 percent confidence interval:
 236.474     NA
sample estimates:
mean of x
 250.0645
```

Apart from labelling and the number of decimals reported, the two functions produce the same result, but BSDA used NA rather than Inf to represent the upper limit of the confidence interval.

Recipe 13-2. Writing Functions That Produce Other Functions
Problem

Recipe 13-1 shows how a function can call other R functions, such as mean and length. Many R functions are actually functions of other functions. When you create an object in a function, that object is *local* to the function. The last value computed in a function will be automatically returned by R, but it is often a good idea to use an explicit return() statement. To illustrate the scoping of the variables and objects in a function, note that we can define x within the function, but that assignment will not affect the value of x in the global environment (our R workspace). Similarly, R functions such as mean and length reside in the global workspace, and can be used within other functions.

The function squareX redefines x as the square of x, and then prints the new vector, but this is local to the function. When you execute the function, the values of x in the global environment are unaffected:

```
squareX <- function(x) {
        x <- x^2
        print(x)
}
> squareX(x)
 [1]  8469.954  5938.378 11432.517  8755.406  8574.257  4004.169  6577.281
 [8]  7978.817  3664.919  8242.086  5112.292  7149.063  7074.971  5202.297
[15]  7303.618  4876.769  5373.086  7577.157  8590.604  7122.608  9195.885
[22] 10393.261  6212.646  4804.475  9342.008  5934.501 10749.049  5390.253
[29]  6821.605  9874.911
> x
 [1]  92.03235  77.06087 106.92295  93.57033  92.59728  63.27850  81.10044
 [8]  89.32423  60.53857  90.78593  71.50029  84.55213  84.11285  72.12695
[15]  85.46121  69.83387  73.30134  87.04686  92.68551  84.39555  95.89518
[22] 101.94735  78.82034  69.31432  96.65406  77.03571 103.67762  73.41834
[29]  82.59301  99.37259
```

Functions can also produce other functions as output, which is useful in a variety of situations. For example, suppose we want to create a function that makes it easier to take various roots of numbers. R has the built-in sqrt function and the exponent operator (^), but perhaps you work with cube roots, and would rather have a function just for that.

Solution

We want to create a function called take.root, which creates a second function called root. We can use the root function to specify the desired root of a given number, as follows. We pass n to the take.root function, which then uses the root function to extract the appropriate root.

```
take.root <- function(n){
        root <- function(x){
                x^(1/n)
        }
        root
}

> sqrRoot <- take.root(2)
> sqrRoot(64)
[1] 8
> cubeRoot <- take.root(3)
> cubeRoot(27)
[1] 3
```

Observe that in our workspace, we now have functions for extracting square roots and cube roots. When you print a function without the ( ), R shows you the function body and the environment in which the function was created. We will discuss R environments and scoping in a little more depth in Recipe 13-3.

```
> ls()
 [1] "chol"      "cholMean"   "cubeRoot"   "LL"           "normalDist"
 [6] "oneSampZ"  "repeated"   "sqrRoot"    "take.root"  "UL"
[11] "x"         "xy"         "zobs"
> sqrRoot
function(x){
x^(1/n)
}
<environment: 0x0239ee44>
```

Recipe 13-3. Writing Functions That Request User Input
Problem

Most of the functions in R produce output that is either printed to the console or passed to some other function. Occasionally, you may want to have a function request input directly from the user. There are many ways to make R interactive, including the web application Shiny by RStudio, as we will discuss in Recipe 13-4. If your aspirations are less ambitious, you can use R functions to request user input from a function.

Solution

Assume that we are interested in obtaining user input for some of the arguments to a function. For example, what if we want a body mass index (BMI) calculator that will prompt the user for his or her height and weight? We could of course pass these as arguments to the function, but we could also ask the user to enter them after being prompted.

Here is our simple BMI function. It prompts the user for his or her height in inches and weight in pounds, and then calculates the BMI. The function also determines the person's classification as underweight, normal, overweight, or obese, according to the standards set by the National Heart, Lung, and Blood Institute.

```
BMI <- function() {
       cat("Please enter your height in inches and weight in pounds:","\n")
       height <- as.numeric(readline("height = "))
       weight <- as.numeric(readline("weight = "))
       bmi <- weight/(height^2)*703
       cat("Your body mass index is:",bmi,"\n")
       if (bmi < 18.5) risk = "Underweight"
       else if (bmi >= 18.5 && bmi <= 24.9) risk = "Normal"
       else if (bmi >= 25 && bmi <= 29.9) risk = "Overweight"
       else risk = "Obese"
       cat("According to the National Heart, Lung, and Blood Institute,","\n")
       cat("your BMI is in the",risk,"category.","\n")
}
```

To use the function, you simply enter BMI() at the R command prompt after copying the function into the working memory. You may recall that the function script can be saved as an R file from the R editor and loaded into the working memory by opening the script, and then selecting all the code and pressing Alt + R to "run" it. Here is how the function looks in operation. The two lines in bold are the points at which the function prompts the user for input:

```
> BMI()
Please enter your height in inches and weight in pounds:
height = 67
weight = 149
Your body mass index is: 23.33415
According to the National Heart, Lung, and Blood Institute,
your BMI is in the Normal category.
```

Recipe 13-4. Taking R to the Web
Problem

Taking R to the Web involves the use of some kind of server, whether you are running it on your local machine or using a web-based server to which you have access. RStudio makes it easy to create web applications using the Shiny package, and we will illustrate that in Recipe 13-4. Shiny is a product of Revolution Analytics, and is provided free of charge, along with the RStudio GUI for R. Many of their other value-added solutions are commercial products. Because Shiny integrates so easily with RStudio, we will illustrate Shiny with RStudio rather than the R Console.

Solution

Every Shiny application must have two scripts. One is the user interface script, and the other is the server script. The server script specifies the input and output and the user interface script specifies what the shiny application will do. For instance, Figure 13-3 shows a basic server.R script in the RStudio interface. Observe the triangle next to Run App in the code (top-left) panel. When you have Shiny installed, you will have this feature available whenever you create a Shiny application. Also notice in the right panel that shiny and shinyapps must both be installed as R packages.

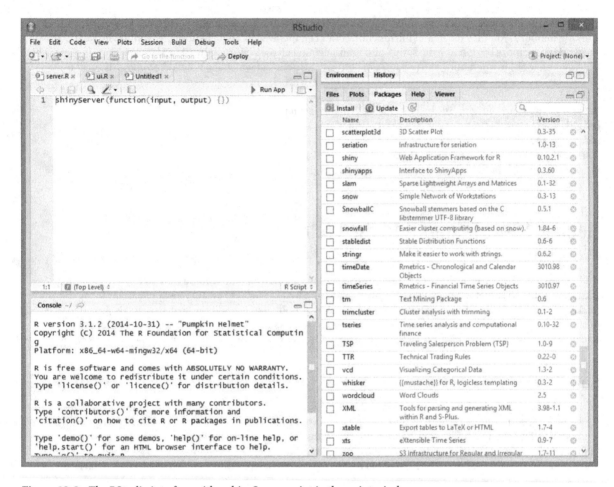

Figure 13-3. *The RStudio interface with a shinyServer script in the script window*

The user interface script contains the commands that make the shiny application. Here is the code for the script ui.R in my RStudio workspace (I will only show the RStudio interface if there is something new to add). One of the nice features of RStudio is that it makes sure your parentheses balance, and using indentation makes the code easier to read. See that you can use HTML-type tags in your Shiny application. Type the following and save it as `ui.R`:

```
shinyUI(fluidPage(
  titlePanel("Check out Shiny"),
  sidebarLayout(
    sidebarPanel( "sidebar panel",
    h3("Shiny is from RStudio")),
    mainPanel("main panel",
      h1("This is a level 1 header!")
  )
  ))
)
```

CHAPTER 13 ■ WRITING REUSABLE FUNCTIONS

Here's what happens when you click Run App (see Figure 13-4). Note that the web application is running on my local host. If you want to look at the app in a browser, you can click Open in Browser. If you want to deploy to app to RStudio's shiny server, you will have to register for a free account.

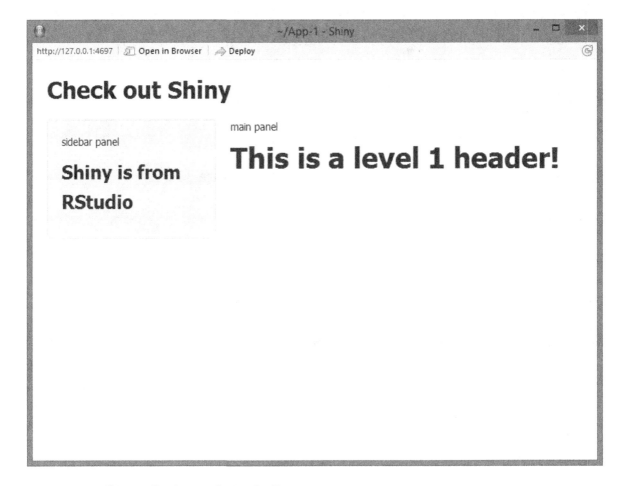

Figure 13-4. *A Shiny application running in RStudio*

After establishing your ShinyApps account, you can host your applications on the ShinyApps server and deploy them by clicking the Deploy button (see Figure 13-5).

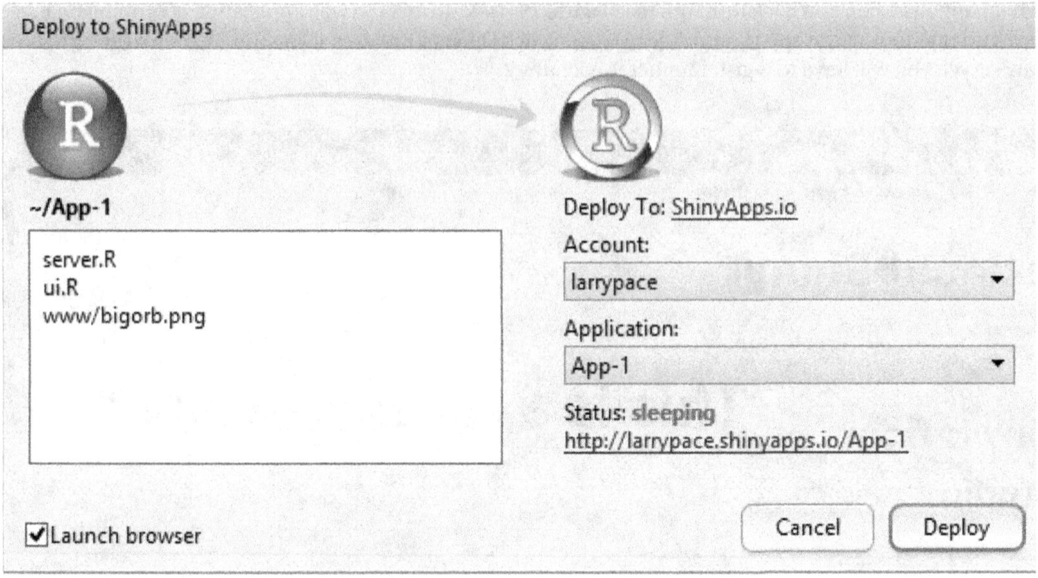

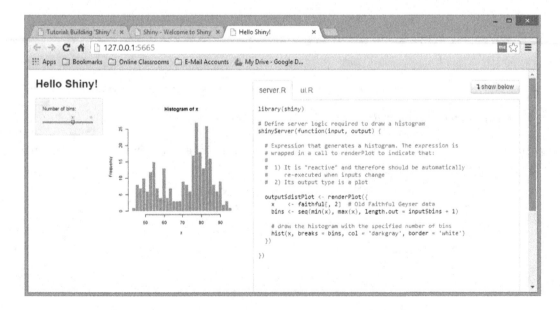

Figure 13-5. Preparing to deploy a Shiny application

The application opens in a browser window with the URL `https://larrypace.shinyapps.io/App-1/`.

You can add interactivity to your apps very easily. For example, here is RStudio's demonstration application of a histogram of the Old Faithful eruption data, with a slider bar for the number of bins. See that every Shiny app must have a server.R script and a ui.R script. You can display the scripts next to the output, or below it if you would rather do so (see Figure 13-6).

Figure 13-6. The "Hello Shiny" demonstration

Here is the ui.R script for the application:

```
library(shiny)

# Define UI for application that draws a histogram
shinyUI(fluidPage(

  # Application title
  titlePanel("Hello Shiny!"),

  # Sidebar with a slider input for the number of bins
  sidebarLayout(
    sidebarPanel(
      sliderInput("bins",
                  "Number of bins:",
                  min = 1,
                  max = 50,
                  value = 30)
    ),

    # Show a plot of the generated distribution
    mainPanel(
      plotOutput("distPlot")
    )
  )
))
```

RStudio provides an excellent tutorial and a gallery of many Shiny apps that you can examine and modify. You can find these resources at http://shiny.rstudio.com. Although you do not have to be running RStudio in order to use Shiny, the two solutions play very well together, as you might expect.

■ ■ ■

## CHAPTER 14

Working with Financial Data

Individuals and organizations work with financial data on a daily basis. While accountants are interested in money from an *accrual* basis, individuals and financial managers in organizations are usually more interested in money from a *cash* basis. That is, they are interested in revenues and expenses only with respect to the actual inflows and outflows of cash. Both individuals and organizations maintain solvency by acquiring the cash flows necessary to satisfy their obligations and to obtain the assets needed to achieve their goals. In this chapter, you will learn how to use R to analyze financial data, and we will primarily adopt the perspective of cash flow rather than accrual.

Recipe 14-1. Getting and Visualizing Financial Data
Problem

Financial data are everywhere, but often may seem inaccessible. You learned in Chapter 7 that you can retrieve financial information such as stock prices from the Yahoo! Finance web site. The Internet provides a wide variety of financial data ripe for the plucking. Those who invest in stock and other equities need to be able to visualize the patterns of change over time, and one of the best ways to do that is to use various charts, such as the bar chart, the candle chart (which bears a superficial resemblance to a boxplot), and the line chart. The quantmod package provides charts—both alone and in combination, as you will observe next.

Solution

You can use R to analyze financial data of various sorts. The quantmod package provides many options for analyzing stock data. Let us use Yahoo! Finance as the source to retrieve Netflix stock information from January 2007 to September 2014. The getSymbols function loads the data directly into the R workspace:

```
> install.packages("quantmod")
> library(quantmod)
> getSymbols("NFLX", src = "yahoo")
[1] "NFLX"
> head(NFLX)
           NFLX.Open NFLX.High NFLX.Low NFLX.Close NFLX.Volume NFLX.Adjusted
2007-01-03     26.00     26.77    25.74      26.61     2348700         26.61
2007-01-04     26.41     26.80    25.10      25.35     2279900         25.35
2007-01-05     25.34     25.34    24.45      24.81     2170100         24.81
2007-01-08     24.82     24.89    23.57      23.83     2620700         23.83
2007-01-09     23.99     24.08    23.52      23.99     1515900         23.99
2007-01-10     23.97     24.25    23.88      24.07     1635500         24.07
```

```
> tail(NFLX)
           NFLX.Open NFLX.High NFLX.Low NFLX.Close NFLX.Volume NFLX.Adjusted
2014-09-23    441.23    448.01   440.50     443.90     1638800        443.90
2014-09-24    444.46    451.66   442.54     450.56     1364500        450.56
2014-09-25    449.90    452.35   442.42     443.49     1405700        443.49
2014-09-26    445.00    450.64   443.89     448.75     1486400        448.75
2014-09-29    444.07    450.54   442.02     449.56     1359500        449.56
2014-09-30    452.30    457.19   449.80     451.18     1781800        451.18
```

We see the traditional *open, high, low, close* (OHLC) data for each trading day, along with the daily trading volume and the adjusted closing price. Let us restrict our data to the second quarter of 2014 in order to get a better view of some of the charting options available in quantmod. It is very easy to subset the data by year and month, as follows. Note the use of the::operator to select the range of dates.

```
> NFLX2Q <- NFLX["2014-04::2014-06"] # April 2014 through June 2014
> head(NFLX2Q)
           NFLX.Open NFLX.High NFLX.Low NFLX.Close NFLX.Volume NFLX.Adjusted
2014-04-01    351.75    365.25   351.74     364.69     3048400        364.69
2014-04-02    365.66    371.05   358.30     362.88     3456300        362.88
2014-04-03    361.33    365.10   350.10     354.69     3140500        354.69
2014-04-04    355.45    356.00   335.88     337.31     4994000        337.31
2014-04-07    340.51    348.19   331.11     338.00     5335900        338.00
2014-04-08    340.05    350.79   338.39     348.89     3680200        348.89
> tail(NFLX2Q)
           NFLX.Open NFLX.High NFLX.Low NFLX.Close NFLX.Volume NFLX.Adjusted
2014-06-23    439.36    441.86   435.55     439.52     1524700        439.52
2014-06-24    438.05    449.94   435.50     436.36     2909600        436.36
2014-06-25    435.00    444.76   433.33     444.21     2250300        444.21
2014-06-26    440.79    442.14   436.75     439.61     2033900        439.61
2014-06-27    438.32    443.19   437.40     442.08     2226800        442.08
```

Figure 14-1 is a candle chart for the selected period. The color scheme chosen (multi.col = TRUE) means the following (taken from the function documentation): a "gray candle" means that the opening price is lower than the closing price on two consecutive trading days; a white one means the opening price is lower than the closing price on the second day, but higher than the closing price on the previous day; red means the opening price is higher than the closing price on the second day, but lower on the previous day; and black means the opening price is higher than the closing price on both trading days (see Figure 14-1). The argument TA = NULL removes the overlay, which defaults to a volume chart.

- gray => Op[t] < Cl[t] and Op[t] < Cl[t-1]

- white => Op[t] < Cl[t] and Op[t] > Cl[t-1]

- red => Op[t] > Cl[t] and Op[t] < Cl[t-1]

- black => Op[t] > Cl[t] and Op[t] > Cl[t-1]

```
> candleChart(NFLX2Q, multi.col = TRUE, theme = "white", TA = NULL)
```

Figure 14-1. *Candle chart for Netflix stock prices, 2ⁿᵈ quarter 2014*

A bar chart is similarly easily implemented (see Figure 14-2). In order to see the white bars, let us replace the "white" theme with the "black" theme:

```
> barChart(NFLX2Q, multi.col = TRUE, theme = "black", bar.type = "hlc", TA = NULL)
```

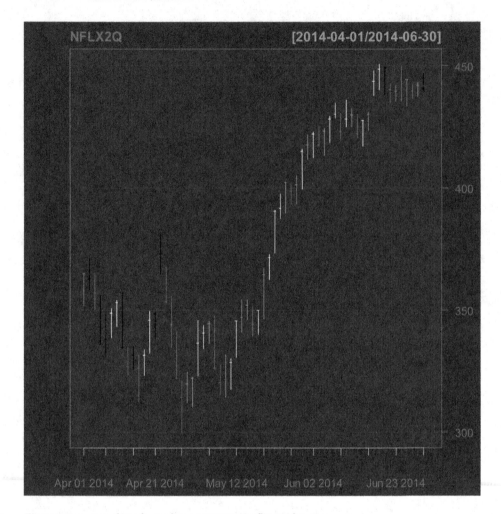

Figure 14-2. *Bar chart for 2nd quarter 2014 Netflix stock prices*

Note that the "bar chart" is not a bar graph in the traditional sense, but a plot for each trading day of the high, low, and closing price of the stock. We use the same color scheme for the bar chart and the candle chart. The line chart (not shown) has similar options. The chartSeries function combines charts as follows (see Figure 14-3).

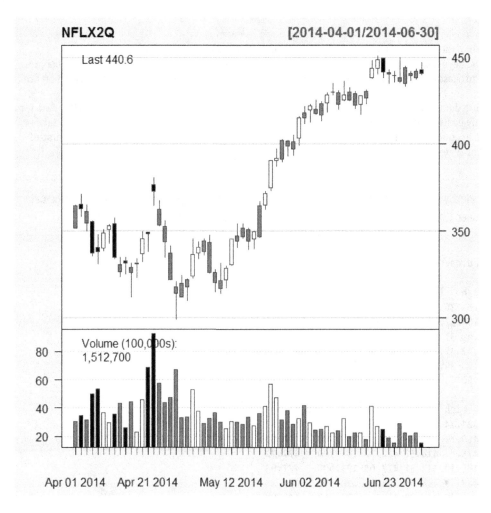

Figure 14-3. *A combined candle and volume chart produced by the* chartSeries *function*

Recipe 14-2. Analyzing Stock Returns
Problem

Making good investment decisions involves understanding risk and reward, and balancing the two. Many factors are at play in the decision to invest, including the time frame, the person's age (or the organization's purpose for the investment), the expected return, and the tax implications of the investment. R provides tools to help a person or an organization make sound investment decisions.

Solution

We continue with our use of the quantmod package, but also explore other packages that focus on financial applications and analysis. Let us examine the performance of Netflix stock and determine if investing in Netflix is a good investment. In the process, we will examine the simple return and the continuously compounded return on our investment in Netflix.

We have the following data (taken from Yahoo! Finance, as discussed earlier), which represent the OHLC data for Netflix stock on a monthly basis from May 2009 to October 2014. We will use the adjusted closing price, which takes splits and dividends into account. Note that most financial analysts would use the end-of-month rather than first-of-month data, but Yahoo! provides the first of month.

■ **Tip** To find the data, simply search for the Netflix historical data and select monthly. Set the dates to select the data from May 1, 2009 to October 1, 2009, and then click Download to Spreadsheet at the bottom of the screen.

```
> NFLX_monthly <- read.csv("NFLX_monthly.csv")
> head(NFLX_monthly)
        date  open  high   low close     vol adjusted
1   5/1/2009 45.25 45.83 36.25 39.42 1963600    39.42
2   6/1/2009 39.99 42.81 37.05 41.34 1859400    41.34
3   7/1/2009 41.58 47.69 37.93 43.94 1762400    43.94
4   8/3/2009 44.35 46.67 42.51 43.63 1073100    43.63
5   9/1/2009 43.19 48.20 39.27 46.17 1326900    46.17
6  10/1/2009 45.75 57.50 44.30 53.45 1496700    53.45
> tail(NFLX_monthly)
         date   open   high    low  close     vol adjusted
61  5/1/2014 324.05 421.74 314.36 417.83 3762400   417.83
62  6/2/2014 419.48 450.82 412.50 440.60 2471000   440.60
63  7/1/2014 456.23 475.87 418.52 422.72 2879700   422.72
64  8/1/2014 421.76 485.30 412.51 477.64 1932600   477.64
65  9/2/2014 478.50 489.29 438.88 451.18 1909100   451.18
66 10/1/2014 448.69 450.49 437.29 449.98 3736900   449.98
```

First, let us examine the growth pattern by using a line chart. To demonstrate R's flexibility, let us use the ggplot2 package to produce the line chart (see Figure 14-4). We can create an index for the number of rows in the Netflix monthly data and use that index as our *x* axis, as follows:

```
> install.packages("ggplot2")
> library(ggplot2)
> n <- nrow(NFLX_monthly)
> index <- 1:n
> index
 [1]  1  2  3  4  5  6  7  8  9 10 11 12 13 14 15 16 17 18 19 20 21 22
[23] 23 24 25 26 27 28 29 30 31 32 33 34 35 36 37 38 39 40 41 42 43 44
[45] 45 46 47 48 49 50 51 52 53 54 55 56 57 58 59 60 61 62 63 64 65 66
> ggplot(NFLX_monthly, aes(x=index,y=adjusted))+geom_line()
```

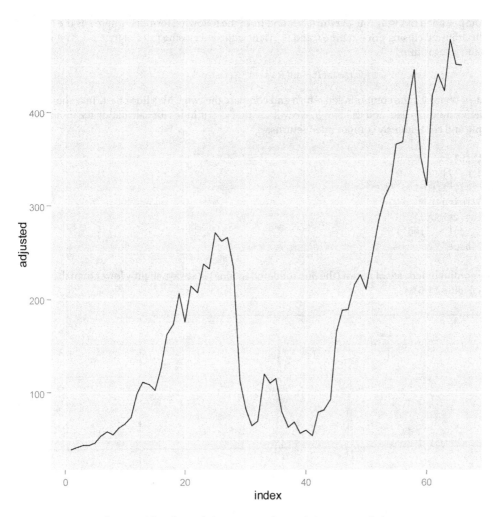

Figure 14-4. *Netflix monthly adjusted closing price from 5/1/2009 to 10/1/2014*

The growth pattern certainly looks encouraging. Let us calculate the simple return for the Netflix stock over the months. We will use the following formula for our calculations:

$$R_t = \frac{P_t - P_{t-1}}{P_{t-1}}$$

where R_t is the simple rate of return, and P_t is the adjusted closing price for month t. We will create a vector of the simple return.

The continuously compounded monthly rate of return is found from the following formula, where r_t is the rate of return at time t, P_t is the adjusted closing price at time t, and P_{t-1} is the adjusted closing price at time $t-1$. The log() function in R finds the natural logarithm:

$$r_t = \ln(P_t) - \ln(P_{t-1})$$

We can also calculate a vector for the compounded return and compare the two with a line chart. Take the number of rows in the Netflix monthly data and call that n. We will use that as our index for calculating the differences for our formulas for simple and continuously compounded returns:

```
> n <- nrow(NFLX_monthly)
> t <- NFLX_monthly[2:n, 7]
> tminus1 <- NFLX_monthly[1:(n-1), 7]
> Rt <- (t - tminus1)/tminus1
> ccret <- log(t) - log(tminus1)
> plot(Rt, type = "l", col = "red")
> lines(ccret, col = "blue")
```

We observe that the compounded rate of return (the line rendered in blue) is always slightly lower than the simple rate of return (see Figure 14-5).

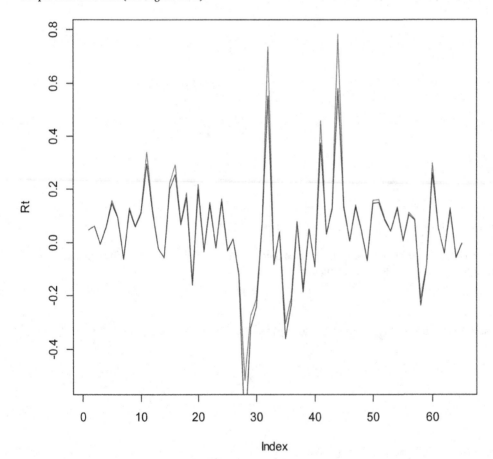

Figure 14-5. *Comparison of simple and continuously compounded returns for Netflix stock*

Now, we have enough information to answer an important question: "If we invested a dollar in Netflix stock on the first trading day of 2013, what would our investment be worth at the end of 2013?" Let's select the daily trading information for Netflix from 2013, and perform the necessary calculations. We will use the Yahoo! Finance web site once again and select the Netflix data for 2013, which you can download in CSV format and then read into R:

```
> nflx_2013 <- read.csv("nflx_2013.csv")
> head(nflx_2013)
       date   open   high   low  clos    vol   adj
1 1/2/2013  95.21  95.81 90.69 92.01 2775900 92.01
2 1/3/2013  91.97  97.92 91.53 96.59 3987500 96.59
3 1/4/2013  96.54  97.71 95.54 95.98 2537300 95.98
4 1/7/2013  96.39 101.75 96.12 99.20 6507200 99.20
5 1/8/2013 100.01 100.99 96.80 97.16 3530700 97.16
6 1/9/2013  97.12  97.95 94.55 95.91 2889000 95.9
```

We can calculate the simple return as before, and then use the cumprod() function to calculate the cumulative product of our investment, which, as we see from inspection of Figure 14-6, would have been a very good one.

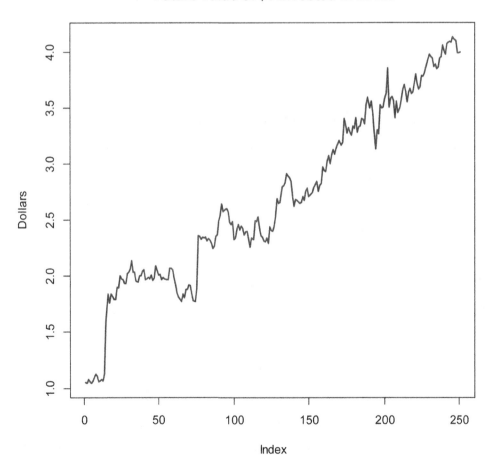

Figure 14-6. *Future value of $1 invested in Netflix stock*

Recipe 14-3. Comparing Stocks
Problem

Most individual and corporate investors would rather have a portfolio of stocks and other investments, instead of investing in a single stock. In Recipe 14-3, you get a brief introduction into the analysis of stock performance of several stocks at the same time. We will keep our examples simple, but in the process, you will learn how to use the same approach with more complex investments.

Solution

We will use three R packages in our solution, the `PerformanceAnalytics` package, the `zoo` package, and the `tseries` package. As you remember, the `quantmod` package requires both the `zoo` package and the `xts` package.

Let us continue with our analysis of Netflix stocks, but add two companies, in this case, Starbucks and Microsoft, to the mix. We want to determine the performance of these three stocks taken together and see if the combination is a good investment. We will get the data from Yahoo! Finance as before, and then get it into the format we need for our analysis. The "zoo" class stands for "Z's ordered observations," and allows the user to work with both regular and irregular time series data. We will retrieve the historical prices for our three stocks from January 2012 to December 2013, and then change the date format to the year and month. Last, we will merge the three sets of prices into one dataset.

```
> install.packages("PerformanceAnalytics")
> library(PerformanceAnalytics);library(zoo);library(tseries);
> NFLX_prices = get.hist.quote(instrument="nflx", start="2012-01-01",end="2013-12-31",
+ quote="AdjClose",provider="yahoo", origin="1970-01-01",compression="m", retclass="zoo")
> SBUX_prices = get.hist.quote(instrument="sbux", start="2012-01-01",end="2013-12-31",
+ quote="AdjClose",provider="yahoo", origin="1970-01-01",compression="m", retclass="zoo")
> MSFT_prices = get.hist.quote(instrument="msft", start="2012-01-01",end="2013-12-31",
+ quote="AdjClose",provider="yahoo", origin="1970-01-01",compression="m", retclass="zoo")
> index(NFLX_prices) = as.yearmon(index(NFLX_prices))
> index(SBUX_prices) = as.yearmon(index(SBUX_prices))
> index(MSFT_prices) = as.yearmon(index(MSFT_prices))
> portfolio_prices <- merge(NFLX_prices,SBUX_prices,MSFT_prices)
> colnames(portfolio_prices) <- c("NFLX","SBUX","MSFT")
> head(portfolio_prices)
           NFLX   SBUX  MSFT
Jan 2012 120.20 46.11 27.30
Feb 2012 110.73 46.89 29.54
Mar 2012 115.04 53.96 30.02
Apr 2012  80.14 55.39 29.80
May 2012  63.44 53.16 27.34
Jun 2012  68.49 51.64 28.66
```

To calculate the continuously compounded returns for our three stocks, we can use the following syntax. Note that the continuously compounded return is the difference in the log prices for each column.

```
> portfolio_returns <- diff(log(portfolio_prices))
> head(portfolio_returns)
                  NFLX         SBUX          MSFT
Feb 2012 -0.08206222   0.01677459   0.078858575
Mar 2012  0.03818509   0.14043860   0.016118549
Apr 2012 -0.36150479   0.02615604  -0.007355433
May 2012 -0.23368053  -0.04109284  -0.086157562
Jun 2012  0.07659317  -0.02900967   0.047151591
Jul 2012 -0.18627153  -0.16352185  -0.037324396
```

We can use the PerformanceAnalytics package's chart.TimeSeries function to create a chart of the returns (see Figure 14-7). We see that the Netflix stock is somewhat more variable than the other two, but also appears to have gained more value overall.

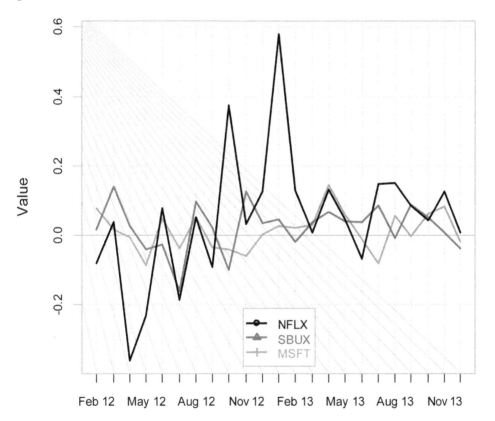

Figure 14-7. *Continuously compounded returns for three stocks*

Calculating and plotting the simple returns will make our analysis more meaningful. The simple returns are calculated for each column in the same way we did before, by dividing the differences between the prices at times t and $t-1$ and dividing by the price at time $t-1$. However, the PerformanceAnalytics package has the difference and lag functions built in to make the calculations easier. The k parameter shows we are using a time lag of one month. The chart.CumReturns function produces a very attractive chart (see Figure 14-8).

```
simple_returns = diff(portfolio_prices)/lag(portfolio_prices, k=-1);
chart.CumReturns(simple_returns, legend.loc="topleft", wealth.index = TRUE, main= "Future Value of
$1 invested")
```

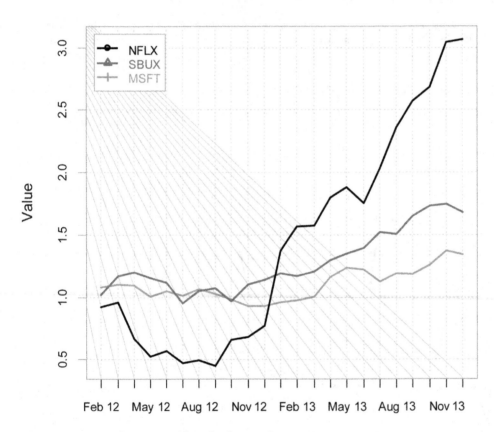

Figure 14-8. *Cumulative return chart for three stocks*

We can also use boxplots to compare the three investments, as follows (see Figure 14-9 for the resulting chart):

```
chart.Boxplot(simple_returns)
```

Return Distribution Comparison

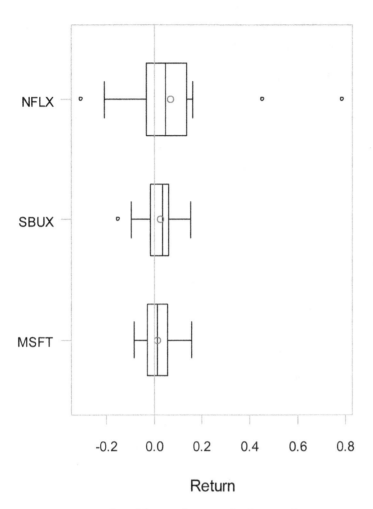

Figure 14-9. *Boxplots of the simple returns for three stocks*

Recipe 14-4. A Brief Introduction to Portfolios
Problem

Now that you have seen how we might analyze stocks and make comparisons of different stocks, important questions still remain. If we want to invest in a diversified combination of stocks or other investments, what will be our expected return? What is our expected risk? To answer these questions, we will briefly explore the constant expected return (CER) model. This model is used quite often in finance for portfolio analysis, Capital Asset Pricing, and option pricing models. The CER model assumes an asset's return over time is independent and normally distributed, with a constant mean and variance. The model allows us to correlate the expected returns on different assets, assuming that the correlations are constant over time. The CER model is equivalent to the measurement error model in statistics.

We state that each asset return is equal to a constant μ_i (the expected return) plus a normally distributed random variable ε_{it} with a mean of zero and a constant variance.

$$r_{it} = \mu_i + \varepsilon_{it}, t = 1, \ldots, T$$

This model serves as the basis for "modern portfolio theory" (MPT), and is widely used, as mentioned earlier. Of course, it is only a model, and like most models, it fails to capture the entirety of what happens in the real world, where markets are not always efficient and investors are not always rational. But it is a good place to start when analyzing different combinations of investments as to their expected return and risk. Just as the assumption of normality is often incorrect in statistics, it is often incorrect in stock performance. To the extent that the assumptions of the CRR model are met, the MPT is an effective representation of the performance of an investment over time, and serves as a benchmark for more complicated models. MPT assumes that investors are averse to risk, and that given two portfolios with the same expected return, investors would choose the less risky one. Generally, assets with higher expected returns are riskier, and each investor must decide how much risk is acceptable. We use the standard deviation of return as an indicator of risk. This is not a particularly good choice if the asset returns are not normally distributed, as discussed briefly.

Solution

Let us simplify our example and include only Netflix and Starbucks stock as our "risky assets." Once you understand the theory, you will be able to apply the same analysis to a more complicated portfolio with more equities. The expected return for a given portfolio is the proportion-weighted combination of the constituent assets' returns. We will use w_i to represent the weighting of a constituent asset (in this case Netflix and Starbucks). Note that the weightings can be negative. Calculate the expected return as follows:

$$E(R_p) = \sum_i w_i E(R_i)$$

where R_p is the return on the portfolio, R_i is the return on asset i, and w_i is the weighting of asset i, which is the proportion of asset i in the portfolio. Let us develop a series of weights for each asset. The weights must add to 1 for each combination.

```
> weights_NFLX <- seq(-1,2,.1)
> weights_SBUX <- 1 - weights_NFLX
> weights_NFLX
 [1] -1.0 -0.9 -0.8 -0.7 -0.6 -0.5 -0.4 -0.3 -0.2 -0.1  0.0  0.1  0.2  0.3  0.4
[16]  0.5  0.6  0.7  0.8  0.9  1.0  1.1  1.2  1.3  1.4  1.5  1.6  1.7  1.8  1.9
[31]  2.0
> weights_SBUX
 [1]  2.0  1.9  1.8  1.7  1.6  1.5  1.4  1.3  1.2  1.1  1.0  0.9  0.8  0.7  0.6
[16]  0.5  0.4  0.3  0.2  0.1  0.0 -0.1 -0.2 -0.3 -0.4 -0.5 -0.6 -0.7 -0.8 -0.9
[31] -1.0
```

Thanks to the excellent instructions of Dr. Eric Zivot of the University of Washington, who teaches computational finance and makes his course available via Coursera, I was able to learn how to develop a graph of my portfolio's relative expected return and relative risk, as follows. First, let us get more current performance data on our two stocks, Netflix and Starbucks. I use the same approach as in Recipe 14-3 to retrieve the monthly adjusted closing prices for the period from January 2013 to September 2014. Merge the two stocks and rename the columns:

```
> NFLX_prices = get.hist.quote(instrument="nflx", start="2013-01-10",end="2014-09-
+ 30",quote="AdjClose",provider="yahoo", origin="1970-01-01",compression="m", retclass="zoo")
```

```
> SBUX_prices = get.hist.quote(instrument="sbux", start="2013-01-10",end="2014-09-
+ 30",quote="AdjClose",provider="yahoo", origin="1970-01-01",compression="m", retclass="zoo")
> colnames(myPortfolio) <- c("NFLX","SBUX")
> head(myPortfolio)
             NFLX  SBUX
2013-01-10 165.24 54.79
2013-02-01 188.08 53.75
2013-03-01 189.28 55.81
2013-04-01 216.07 59.62
2013-05-01 226.25 62.08
2013-06-03 211.09 64.41
> tail(myPortfolio)
             NFLX  SBUX
2014-04-01 322.04 70.13
2014-05-01 417.83 72.99
2014-06-02 440.60 77.12
2014-07-01 422.72 77.42
2014-08-01 477.64 77.81
2014-09-02 451.18 75.46

> myPortfolio_returns <- diff(myPortfolio)/lag(myPortfolio, k = -1)
```

Now, assuming the return is constant over months, we calculate the annualized mean, variance, standard deviation, covariance matrix, covariance of the two stocks, and correlation of the two stocks, as follows. We will use these statistics to calculate the expected return (mean) and the expected risk (the standard deviation) for each combination of weightings defined earlier.

```
> mean_annual <- apply(myPortfolio,2,mean)*12
> var_annual <- apply(myPortfolio,2,var)*12
> sd_annual <- sqrt(var_annual)
> covMat_annual <- cov(myPortfolio)*12
> covHat_annual <- cov(myPortfolio)[1,2]*12
> corr_annual <- cor(myPortfolio)[1,2]
```

Let us now calculate the mean expected performance for each portfolio, and then calculate the variance and standard deviation for each portfolio, as follows. The variance of a portfolio relies on the previously calculated statistics. The formula is

$$\sigma_p^2 = w_A^2\sigma_A^2 + w_B^2\sigma_B^2 + 2w_Aw_B\sigma_{AB}$$

where the w's are the portfolio weights, σ^2 is the variance, and σ_{AB} is the covariance of the two stocks. Here is the vector of means for the different portfolios:

```
> mean_portfolio <- weights_NFLX*mean_annual["NFLX"] + weights_SBUX*mean_annual["SBUX"]
> mean_portfolio
 [1] -0.27792745 -0.22954062 -0.18115378 -0.13276695 -0.08438012 -0.03599329
 [7]  0.01239354  0.06078037  0.10916721  0.15755404  0.20594087  0.25432770
[13]  0.30271453  0.35110137  0.39948820  0.44787503  0.49626186  0.54464869
[19]  0.59303552  0.64142236  0.68980919  0.73819602  0.78658285  0.83496968
[25]  0.88335651  0.93174335  0.98013018  1.02851701  1.07690384  1.12529067
[31]  1.17367751
```

We can now calculate the variance and standard deviation of our portfolios in the following way:

```
#var_portfolio =  weights_NFLX^2 * var_NFLX + weights_SBUX^2 * var_SBUX + 2 * weights_NFLX*
weights_SBUX*cov_NFLX_SBUX
#sd_portfolio = sqrt(var_portfolio)
```

At last we can plot our different portfolios. An efficient portfolio is one that produces a reasonable expected return with a reasonable risk level. See Figure 14-10 for the plot, which is in the shape of a hyperbola.

```
plot(sd_portfolio, mean_portfolio, type="b", pch=16, ylim=c(0, max(mean_portfolio)), xlim=c(0,
+ max(sd_portfolio)), xlab=expression(sigma[p]), ylab=expression(mu[p]), col=c(rep("green", 11), +
rep("red", 20)))
text(x=sd_NFLX, y=mean_NFLX, labels="Netflix", pos=4)
text(x=sd_SBUX, y=mean_SBUX, labels="Starbucks", pos=2)
```

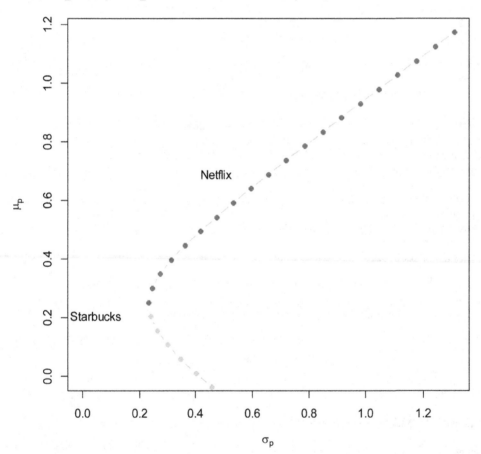

Figure 14-10. *Plotting the different portfolios*

Dr. Zivot has provided an R function for his computational finance class at the University of Washington that calculates the global minimum variance portfolio, which provides the lowest possible volatility for the underlying assets. This and Dr. Zivot's other R functions are available at http://faculty.washington.edu/ezivot/econ424/portfolio.r.

We will find the global minimum variance portfolio and plot its position on a slightly modified version of our previous graph (see Figure 14-11). The globalMin.portfolio() function requires only the input of the expected returns (the means_annual in our case) and the covariance matrix (covMat_annual in this case).

```
> GMV <- globalMin.portfolio(mean_annual, covMat_annual)
> GMV$sd
[1] 0.1556587
> GMV$er
[1] 0.2570875

> plot(sd_portfolio, mean_portfolio, type="b", pch=16, ylim=c(0, max(mean_portfolio)),
xlim=c(0, max(sd_portfolio)), xlab=expression(sigma[p]), ylab=expression(mu[p]),
col=c(rep("green", 11), rep("red", 20)))
> text(x=sd_NFLX, y=mean_NFLX, labels="NFLX", pos=4)
> text(x=sd_SBUX, y=mean_SBUX, labels="SBUX", pos=4)
> text(x=GMV$sd,y=GMV$er,labels="Gobal Min", pos = 2)
```

Figure 14-11. *Plotting the location of the global minimum variance portfolio*

## CHAPTER 15

Dealing with Big Data

Datasets are big and getting bigger, as has been mentioned previously in this book. People talk of the five Vs of big data, which include *volume, variety, velocity, veracity* (or lack thereof), and *value* (or lack thereof). The value gleaned from big data may be short-lived, simply because of the velocity with which the data can change. To deal with bigger data, we need faster processing for larger datasets. To accomplish this, we can use parallel processing, speed up the operations of our processing, use more efficient algorithms, or some combination of these approaches. In this chapter, you will learn how to use the R packages to perform parallel processing, how to extend (and speed up) the capabilities of the traditional R data frame by using data tables instead, and how to speed up computations in R by using compiled code from C++ as well as by preallocating result objects.

Recipe 15-1. Parallel R
Problem

Computing requirements increase when datasets become larger and computations become more intensive. One way to increase computing speed is to distribute the computing tasks to two or more processing units. This is the basic idea behind dual-core and multi-core processors. To the extent that the processing task can be divided among different processors without creating interference or requiring extensive communication among the processors, parallel processing can be more efficient than sequential processing. The ideal is what is known as "embarrassingly parallel" or "perfectly parallel" processing, in which little or no effort is required to separate the problem into a number of parallel problems. In the ideal situation, there is no overhead associated with parallel processing, so a single sequential process that takes 20 seconds to run should take 5 seconds to run with four parallel processing units.

Parallel processing can be accomplished in different ways. We can use a multi-core processor on a single machine, combine multiple workstations (as in a computer lab), or use computer clusters—as in a supercomputer. In R, parallel processing works by establishing a *cluster* of processing units, each of which is an instance of R, one of which one is the primary ("master processor"), and the others are the secondary ("worker processors").

Parallel R involves the creation of a cluster of R processes on the same or different machines. The master processor is the one in which we create the cluster, and the worker processors wait for and process the parts of the processing task distributed to them. The master processor consolidates the results from the separate workers and produces the final output. In the real world, unlike the ideal one, there is overhead associated with parallel processing, so some tasks do not benefit from parallel processing because of overhead, while computation-intensive processes can benefit substantially.

Solution

Parallel computing can be either implicit or explicit, depending on the level at which the hardware and software come together. For example, the pnmath package by Luke Tierney uses the OpenMP parallel processing directives of recent compilers for implicit parallelism by replacing some internal R functions with substitute functions that can make use of multiple cores without any explicit requests from the user. There are several other such packages that provide implicit parallelism, but our discussion will focus on explicit parallelism.

Unlike implicit parallelism, explicit parallelism requires a communications layer. The first R package providing this capability was rpvm by Na Li and Tony Rossini. This package used the PVM (Parallel Virtual Machine) standard and libraries. More recently, the MPI (Message Passing Interface) has become more popular. It is supported in R via the Rmpi package written by Hao Yu. There are other standards, including NWS (NetWorkSpaces), and sockets. A *socket cluster* is a set of copies of R running in parallel, often on the same machine, though it is also possible to specify the host names on which the worker processes are to be executed as well. If the sockets are to be formed on the same machine, a positive integer specifies the number of socket clusters. For example, if we use the snow package written by Luke Tierney, and we have a quad-core machine, we can set up four clusters as follows. The original R process, that is, the one in which we are typing the commands, is the master (or manager) process. We are setting up four new worker processes, one on each of our processors. Each process is a separate R session, so we are creating the ability to run multiple R sessions at the same time (or in parallel). See that each of these clusters is running on localhost, the standard network name for the local machine.

```
> install.packages("snow")
> library(snow)
> cl <- makeCluster(type = "SOCK", 4)
> cl
[[1]]
$con
        description          class           mode            text
"<-Larry2in1:11720"    "sockconn"         "a+b"        "binary"
            opened       can read       can write
          "opened"         "yes"          "yes"

$host
[1] "localhost"

$rank
[1] 1

attr(,"class")
[1] "SOCKnode"

[[2]]
$con
        description          class           mode            text
"<-Larry2in1:11720"    "sockconn"         "a+b"        "binary"
            opened       can read       can write
          "opened"         "yes"          "yes"

$host
[1] "localhost"

$rank
[1] 2
```

```
attr(,"class")
[1] "SOCKnode"

[[3]]
$con
        description              class               mode               text
"<-Larry2in1:11720"        "sockconn"              "a+b"           "binary"
                opened          can read          can write
             "opened"              "yes"              "yes"

$host
[1] "localhost"

$rank
[1] 3

attr(,"class")
[1] "SOCKnode"

[[4]]
$con
        description              class               mode               text
"<-Larry2in1:11720"        "sockconn"              "a+b"           "binary"
                opened          can read          can write
             "opened"              "yes"              "yes"

$host
[1] "localhost"

$rank
[1] 4

attr(,"class")
[1] "SOCKnode"

attr(,"class")
[1] "SOCKcluster" "cluster"
```

The acronym snow stands for Simple Network of Workstations. The snow package uses a "scatter and gather" approach in which the manager process distributes "chunks" of the task to the worker processors and then gathers the results and combines them. In addition to socket clusters, snow supports MPI, NWS, and PVM, as discussed earlier. There are now many other R packages that support either explicit or implicit parallelism. The snowfall package written by Jochen Knaus is built on snow and uses its network and cluster capabilities. The snowfall package provides the ability to connect directly to R-specific clusters via the sfCluster function and gives the user additional flexibility, such as command-line arguments for configuration. Tierney has more recently authored the parallel package, and there are several additional packages for parallel computing available, including the doParallel and doSnow packages written by Steve Weston at Revolution Analytics, who is also the author of the foreach package. The foreach package provides an idiom for iterating over the elements in a collection without the use of explicit loop counting. The main intention is to produce a return value, and in that sense, foreach is similar to lapply, which we have discussed previously. The doParallel package provides support for the foreach package and acts as an interface between foreach and parallel.

Among the many choices available, we will begin with the use of the doParallel package. The doParallel package requires the user to register a parallel "backend." A backend in this context is the explicit notification to the %dopar% operator that there is a cluster of worker processes available to it. If the backend is not registered, the foreach package will run the tasks sequentially rather than in parallel. After installing doParallel and loading it, we create a cluster, register the backend, and then use foreach to distribute the parts of the task to the "worker" processors. The %dopar% operator makes the processing parallel, while the %do% operator makes it sequential. First, let us make sure we have the required package and load it:

```
> install.packages("doParallel")
> library(doParallel)
Loading required package: foreach
foreach: simple, scalable parallel programming from Revolution Analytics
Use Revolution R for scalability, fault tolerance and more.
http://www.revolutionanalytics.com
Loading required package: iterators
Loading required package: parallel"
```

Observe that the doParallel package requires the foreach, iterators, and parallel packages. Let us begin with a very simple "proof of concept" demonstration before getting down to more serious computations. Simply loading the doParallel package is not sufficient. If you load the package and then use the %dopar% operator, the processing will still be sequential, as shown next. You must register the backend once per session, in which case the clusters you have created will work in parallel. If the backend is not registered, the processing will continue to be sequential, though the warning to that effect will appear only once per session. As we discussed earlier, we must use makeCluster with a positive integer (or with a list of host names) to specify the number of worker processes.

```
> cluster <- makeCluster(2)
> foreach(i=1:5) %dopar% i^(i+1)
[[1]]
[1] 1

[[2]]
[1] 8

[[3]]
[1] 81

[[4]]
[1] 1024

[[5]]
[1] 15625

Warning message:
executing %dopar% sequentially: no parallel backend registered
```

So, we must register the backend only once per session. We will now have parallel processing with our proof of concept example:

```
> clusters <- makeCluster(2)
> registerDoParallel(clusters)
> foreach(i=1:5) %dopar% i^(i+1)
```

```
[[1]]
[1] 1

[[2]]
[1] 8

[[3]]
[1] 81

[[4]]
[1] 1024

[[5]]
[1] 15625
```

The defaults in doParallel are reasonable, and will work for most users most of the time. To find out how many worker processes you have access to, you can use the getDoParWorkers function:

```
> getDoParWorkers()
[1] 2
```

You can also use doParallel with multi-core functionality, but you will be limited to the number of physical cores available to you. For example, the Toshiba laptop computer on which I am working at the moment has a dual-core processor. See what happens when I try to use more than two processors:

```
> options(cores = 2)
> getDoParWorkers()
[1] 2
> options(cores = 4)
> getDoParWorkers()
[1] 2
```

Having changed offices in the middle of the day, I am now working on a machine with a quad-core processor. Note that I can now use all four cores:

```
> options(cores = 4)
> getDoParWorkers()
[1] 4
```

Now that you understand the basics, we will use a less trivial example. As those who have worked with large datasets or computer-intensive calculations well know, R is not a speed demon compared to compiled languages. The use of vectorized operations and functions yields speed efficiencies, but there are times when R is slow compared to other languages. Remember that R is interpreted, not compiled. You can connect R with compiled code from other languages, such as C++, a subject upon which we will touch very lightly in Recipe 15-3. Meanwhile, let us examine the use of parallel R and determine what kinds of speed improvements we might experience.

Adapting Weston's "toy example" in the doParallel documentation, we will use two clusters to perform a bootstrapping problem, and compare the results of sequential and parallel processing by using the system.time function. We will run 10,000 bootstrap iterations in parallel and compare the results to the same task performed sequentially. The system.time function will give us the elapsed time, and the [3] index will return it at the end of the processing. Here is our code. First, let us use the iris dataset and select 100 rows of data with the two species *versicolor* and *virginica*. We will use the lm function to predict sepal length using the species as the predictor.

```
> x <- iris[which(iris[,5] != "setosa"), c(1, 5)]
> head(x)
   Sepal.Length    Species
51          7.0 versicolor
52          6.4 versicolor
53          6.9 versicolor
54          5.5 versicolor
55          6.5 versicolor
56          5.7 versicolor
> tail(x)
    Sepal.Length   Species
145          6.7 virginica
146          6.7 virginica
147          6.3 virginica
148          6.5 virginica
149          6.2 virginica
150          5.9 virginica
```

Now, we can run the process in parallel using two clusters:

```
> clusters <- makeCluster(2)
> registerDoParallel(clusters)
trials <- 10000
parallelTime <- system.time({
        r <- foreach(icount(trials), .combine=cbind) %dopar% {
                ind <- sample(100, 100, replace = TRUE)
        result1 <- lm(x[ind,1]~x[ind,2])
        coefficients(result1)
        }
})[3]
parallelTime
elapsed
  25.59
```

By changing %dopar% to %do%, we can process sequentially for comparison purposes:

```
> trials <- 10000
> sequentialTime <- system.time({
        r <- foreach(icount(trials), .combine=cbind) %do% {
        ind <- sample(100, 100, replace = TRUE)
        result1 <- lm(x[ind,1]~x[ind,2])
        coefficients(result1)
        }
 })[3]
sequentialTime
elapsed
  33.89
```

Parallel processing with two clusters was about 24.5% faster than sequential processing in this case. Remember that this is simply the performance of my particular machine and its dual-core processor. Your results are very likely to be different.

The doSNOW package works in a similar way to doParallel, allowing the user to register the snow parallel backend for the foreach package. In this case, we will specify socket clusters. In this particular instance, doSNOW produced slightly faster parallel processing than doParallel:

```
clusters <- makeCluster(2, type = type = "SOCK")
registerDoSNOW(clusters)
trials <- 10000
parallelTime <- system.time({
        r <- foreach(icount(trials), .combine=cbind) %dopar% {
                ind <- sample(100, 100, replace = TRUE)
        result1 <- lm(x[ind,1]~x[ind,2])
        coefficients(result1)
        }
})[3]
> parallelTime
elapsed
  22.36
```

Recipe 15-2. Using Data Tables
Problem

One of the drawbacks of R is that the traditional data frame becomes cumbersome and slow with very large datasets. The data.table package provides additional features to the data frame that make it faster and easier to use. The data table created by the function is also a data frame, so all the things you can do with data frames work with tables as well, with a few exceptions as noted next.

Solution

Those familiar with creating data frames and subsetting them, as we have discussed in this book, will find the data table to be intuitive. We create data tables in the same way that we create data frames. Observe that the data table uses colons to separate the row numbers from the first column of data.

```
> install.packages("data.table")
> library(data.table)
> DataFrame <- data.frame(x = c("b","b","b","a","a"),v = rnorm(5))
> DataFrame
  x          v
1 b -1.1006306
2 b -1.3294956
3 b -0.2006320
4 a  1.6582277
5 a  0.1527988
> DataTable <- data.table(x = c("b","b","b","a","a"),v = rnorm(5))
> DataTable
   x          v
1: b -0.1329364
2: b  0.6700842
3: b  0.4488872
4: a  0.2374386
5: a  1.5203938
```

Instead of creating a data table from scratch as before, you can convert any existing data frame to a data table. To illustrate, we will download the SPSS version of the General Satisfaction Survey discussed earlier and then import it using the Hmisc package. The result will be the expected data frame, which we can then convert to a data table.

```
> install.packages("Hmisc")
> library(Hmisc)
> GSSdataframe <- spss.get("GSS2012Merged_R5.sav", use.value.labels = TRUE)
```

As we did before, we receive warning messages while importing the data into R because of the treatment of missing values in the GSS data. You can look at the GSS data in a spreadsheet-like format by using the View() function. Although not big by the standards of today, our dataset is large enough, with 4,820 rows and 1,069 columns. We convert the data frame to a data table simply by using the data.table function. The data table is using 22 MB of storage. By contrast, the following CARS table is a data table created from the cars data that ship with R, and it uses only 1 MB of storage.

```
> CARS <- data.table(cars)
> View(GSSdataframe)
> nrow(GSSdataframe)
[1] 4820
> ncol(GSSdataframe)
[1] 1069
> GSSdataTable <- data.table(GSSdataframe)
> tables()
Loading required package: portfolio
Loading required package: nlme
        NAME          NROW  NCOL MB
[1,] CARS              50     2  1
[2,] DataTable          5     2  1
[3,] GSSdataTable   4,820 1,069 22
        COLS
[1,] speed,dist
[2,] x,v
[3,] year,id,wrkstat,hrs1,hrs2,evwork,wrkslf,wrkgovt,OCC10,INDUS10,marital,divorce,wi
        KEY
[1,]
[2,]
[3,]
Total: 24MB
```

Instead of dealing with row names as in data frames, we work with *keys* in data tables. You can set a key by using setkey(). The key can be one or more columns in the data table. Each data table may have only one key. Let us experiment with the DataTable object stored in the working memory. The table is now sorted by the key. When we run the tables() command again, we see that DataTable indeed has a key, while we have not yet assigned keys to the CARS or the GSS data tables.

```
> setkey(DataTable, x)
> DataTable
    x         v
1: a  0.2374386
2: a  1.5203938
3: b -0.1329364
```

```
4: b  0.6700842
5: b  0.4488872
> tables()
      NAME        NROW  NCOL MB
[1,] CARS           50    2  1
[2,] DataTable       5    2  1
[3,] GSSdataTable 4,820 1,069 22
      COLS
[1,] speed,dist
[2,] x,v
[3,] year,id,wrkstat,hrs1,hrs2,evwork,wrkslf,wrkgovt,OCC10,INDUS10,marital,divorce,wi
      KEY
[1,]
[2,] x
[3,]
Total: 24MB
```

Now, we can use the key to retrieve specific rows of data. For example, we can use the key to retrieve only the rows in the data table with the value of "b" in the x column:

```
> DataTable["b",]
   x        v
1: b -0.1329364
2: b  0.6700842
3: b  0.4488872
```

The comma shown in the previous command is optional:

```
> DataTable["b"]
   x        v
1: b -0.1329364
2: b  0.6700842
3: b  0.4488872
```

Let us examine the speed improvements we can get with the use of keys in a data table as compared to the use of comparison operators in a data frame. To make the illustration more meaningful, we will simulate some data. Building on the example in the documentation for the data.table package, we will create a data frame with approximately 10 million rows and 676 groups. We generate two columns—one with capital letters and one with lowercase letters—to create the groups, and then create a vector of z scores using rnorm():

```
grpsize <- ceiling(1e7/26^2)
tt <- system.time(DF <- data.frame(
x = rep(LETTERS, each=26*grpsize),
y = rep(letters, each=grpsize),
z = rnorm(grpsize*26^2),
stringsAsFactors = FALSE))[3]
> tt
elapsed
   1.89
```

Now, let's extract a group from the data frame, the cases in which the value of x is "L" and the value of y is "p". We use the standard approach to subsetting the data, which includes comparison operators for the two variables. The time is not terrible. It took my system 2.14 seconds to extract the 14,793 records that match the criteria. R is using vectorized searching here:

```
> tt <- system.time(answer1 <- DF[DF$x == "L" & DF$y == "p",])[3]
> tt
elapsed
  2.14
```

For comparison's sake, let's convert the data frame to a data table and use the key to extract the same information as before:

```
> DT <- as.data.table(DF)
> setkey(DT, x, y)
> tt <- system.time(answer2 <- DT[list("L", "p")])[3]
> tt
elapsed
  0.01
> head(answer1)
        x y            z
4452694 L p   0.48961111
4452695 L p  -0.09609168
4452696 L p   0.26847516
4452697 L p   1.11715638
4452698 L p  -0.77547270
4452699 L p   0.19132284
> head(answer2)
    x y            z
1:  L p   0.48961111
2:  L p  -0.09609168
3:  L p   0.26847516
4:  L p   1.11715638
5:  L p  -0.77547270
6:  L p   0.19132284

> identical(answer1$z, answer2$z)
[1] TRUE
```

We can use the identical() function to verify that the two answers are equivalent. Both sets of retrieved information are the same. The extraction of exactly the same information from the data.table was more than 200 times faster than the traditional data frame approach. The reason the data table was so much faster is that with the table we used a binary search rather than a vector search, as we did with the data frame. Here is an important lesson, paraphrased from a footnote in the data.table documentation: If you use parallel techniques to run vectorized operations, your efficiency may be much lower than simply using a data table.

What we just did was to "join" DT to the 1 row, 2 column table returned by list("L", "p"). Because this is a frequent activity, there is an alias, namely .(), which is a "dot" command followed by parentheses enclosing the search criteria for the key (in this case, the x and y columns of the table).

```
> identical(DT[list("L", "p"),], DT[.("L", "p"),])
[1] TRUE
```

The data table allows us to do very fast grouping as well. The second argument to the data table may be one or more expressions whose arguments are column names (without quotes). The third argument is a by expression. Let's compare the use of the `tapply()` function to the use of the by in `data.table`. We see something on the order of a tenfold increase in speed.

```
> tt <- system.time(total1 <- tapply(DT$z, DT$x, sum))[3]; tt
elapsed
   1.65
> tt <- system.time(total2 <- DT[,sum(z), by = x])[3]; tt
elapsed
   0.17
```

We can also group on two columns should we need to do so. This is also much faster than using `tapply()`. I will leave the performance comparisons to the interested reader.

```
> ttt <- system.time(total3 <- DT[,sum(z), by = "x,y"]); ttt
   user   system elapsed
   0.18     0.03    0.22
```

Recipe 15-3. Compiled Code and Preallocation
Problem

Working with big data is becoming more and more the norm. There are many suggestions for increasing the speed of processing, input and output operation, and data management. Some of these are obvious, such as upgrading hardware, upgrading software, parallel computing (where it makes sense), using data tables as illustrated in Recipe 15-2, and using other languages besides R for time-critical functions. We will touch briefly in Recipe 15-3 on the last item, using another language. Because C++ is compiled, we can take advantage of that fact and use compiled code to speed up computations. We will also use another quick tip to make your code execute more quickly, which is as simple as preallocating a result object before filling it.

Solution

Although we have been using `system.time` to determine the elapsed time for executing our code, we will use the `microbenchmark` package for more precise comparisons in Recipe 15-3. Let us download and install the `Rcpp` package as well, which will provide a seamless integration of R and C++. We can write C++ equivalents to R functions, compile them, and then use them as if they were R functions. Unlike R, C++ performs looping very efficiently, and a compiled C++ object in R can be much more efficient than even vectorized R functions.

The first step for a Windows user is to install `Rtools`, if you do not have them, and to make sure that the `Rtools` are in the path so that R can find them. This will enable you to compile C++ code from within the `Rcpp` package. The `Rtools` are available from CRAN at `http://cran.r-project.org/bin/windows/Rtools/`. Follow the installation instructions, and if you like, you can have the installer add the `Rtools` to your path rather than editing the path manually. If you have a Mac system, you will need Xcode to compile C++ code. Users of Linux systems will have a C++ compiler with their Linux distribution.

After installing `Rcpp` and having the `Rtools` available, you will be able to write small segments of C++ code that can be used in place of R functions, and which work in basically the same way, except that the C++ code will be compiled rather than interpreted. As C++ is quite a "big" language, we will focus on just a few examples to get you started. If this is a subject you find interesting or useful in your work, you will find Dirk Eddelbuettel's book, *Seamless R and C++ Integration with Rcpp* (Springer, 2013), to be very helpful. As a couple of pointers (or reminders), R is not "typed," but C++ is a strongly typed language, and you must declare the type of a variable before you can use it.

Also, unlike R, iterations in C++ begin with 0, not with 1. The for loop in C++ is different from the one in R, and unlike R, C++'s operations are not vectorized.

In addition to compiling C++ code from files saved with the *.cpp extension, you can use C++ inline in R for small code segments. As this is the simplest way to get started, we will look at a couple of examples. The more useful approach will be saving the C++ source file and then "sourcing it" in R using the sourceCpp function in Rcpp.

Adapting an example from Eddelbuettel, let's compare an R function to calculate the nth number in a Fibonacci sequence to an inline function we create using Rcpp and C++. Recall that the Fibonacci sequence begins with 0, 1, 1, 2, 3, 5, 8, 13, 21, 34, 55. Each successive number is the sum of the previous two numbers. A rather straightforward R function can be written to find the Fibonacci number for any position in the sequence. The Fibonacci sequence can be defined mathematically as follows, with seed values $F_0 = 0$ and $F_1 = 1$.

$$F_n = F_{n-1} + F_{n-2}$$

We can program this in R with a relatively simple function:

```
fibR <- function(n) {
        if (n == 0) return(0)
        if (n == 1) return(1)
        return(fibR(n-1)+fibR(n-2))
}
```

To make a C++ function work in R, we can use Rcpp as follows. In this particular case, the C++ function is almost a literal translation of the R code. The question is whether the C++ function is faster than the R function. We will answer that question using the microbenchmark package after verifying that the two functions produce the same results. Simply execute the following script to create and compile the function fibCpp. Note that lines of C++ code must be terminated by a semicolon (;):

```
library(Rcpp)
cppFunction('int fibCpp(int n) {
        if(n == 0) return(0);
        if(n == 1) return(1);
        return(fibCpp(n -1) + fibCpp(n - 2));
}
' )
```

The functions produce the same result:

```
> fibCpp(20)
[1] 6765
> fibR(20)
[1] 6765
```

However, the compiled C++ function is much faster, as the microbenchmark indicates:

```
> library(microbenchmark)
> microbenchmark(fibCpp(20), fibR(20))
Unit: microseconds
       expr       min        lq        mean     median          uq         max
 fibCpp(20)     42.92    44.319    51.75065    49.4505      57.848      99.367
   fibR(20)  72050.87 72297.183 72963.34808 72483.5550 72632.372 109907.356
```

```
neval
 100
 100
```

Be aware that although the newly created function is in your workspace for repeated use in the current session, the compiled code is not saved, and you will need to compile the function again to use it in a new session. As I mentioned, it is more useful to write the C++ function and save it in a *.cpp file and source it when you need it. Here is an example. We will take a commonly used R function and produce a C++ version of it. In this case, we will see how much of a savings we might get by calculating the variance of a dataset with C++ as compared to using R's built-in var function. You can use the R Editor or any other text editor to write the following code and save it as a *.cpp file. I chose the self-explanatory name varC.cpp. The function uses a two-pass algorithm to find the mean and then the variance using the definitional formula for the variance:

$$s_x^2 = \frac{\sum(x - \bar{x})^2}{n-1}$$

As before, we must declare the type for x, n, and the other variables in our function. The use of differing forms of commenting allows us to embed R code, which the compiler sees as a comment and ignores. The size command in C++ determines the length of the x vector. In C++, it is common to declare a variable's type and assign an initial value to it at the same time. Here are the contents of the varC.cpp file, which I created using the R Editor.

```cpp
#include <Rcpp.h>
using namespace Rcpp;

// [[Rcpp::export]]
double varC(NumericVector x) {
  int n = x.size();
  double sum1 = 0;
  double sum2 = 0;
  double mean = 0;
  double var = 0;

  for(int i = 0; i < n; ++i) {
    sum1 += x[i];
  }
  mean = sum1 / n;

  for(int i = 0; i < n; ++i) {
    sum2 += (x[i]-mean)*(x[i]-mean);
  }
  var = sum2 / (n-1);
  return var;
}

/*** R
library(microbenchmark)
x <- rnorm(1e6)
microbenchmark(
  var(x),
  varC(x)
)
*/
```

To compile and execute this function, use the source.Cpp function in Rcpp. We create a vector of 1,000,000 random normal deviates, and see how long it takes our C++ function to find the variance as compared to the var function in R. Although the time savings are not as dramatic in this case, we still achieve a decent improvement, even though the R function is already vectorized.

```
> library(Rcpp)
> sourceCpp("varC.cpp")

> library(microbenchmark)

> x <- rnorm(1e6)

> microbenchmark(
+    var(x),
+    varC(x)
+ )
Unit: milliseconds
    expr      min       lq     mean   median       uq      max neval
  var(x) 6.947759 6.992777 7.041822 7.020535 7.077449 7.377417   100
 varC(x) 3.140090 3.179044 3.200733 3.190008 3.207968 3.311300   100
```

As a last illustration of making code more efficient, note the difference between the time required to fill a result object when you preallocate it as compared to simply creating the object. This is a way to save more time, especially when working with loops (which are sometimes unavoidable). Here, we create a loop that produces 100,000 random normal deviates. In the first case, we simply declare x1 to be a numeric vector and then fill it using the loop. In the second case, we preallocate the vector and then fill it. Notice the substantial speed difference the preallocation makes.

```
> timer <- system.time({
+ x1 <- numeric()
+ for(i in 1:100000) x1[i] = rnorm(1)
+ })[3]
> timer
elapsed
  10.03
> timer <- system.time({
+ x1 <- numeric(100000)
+ for(i in 1:100000) x1[i] = rnorm(1)
+ })[3]
> timer
elapsed
   0.52
```

We have scratched the surface of the very powerful and flexible Rcpp package, but I hope the interested reader has enough information from this recipe to get started and to continue to learn how to work with the package.

CHAPTER 16

■ ■ ■

Mining the Gold in Data and Text

Part of the Big Data revolution is the rapid growth of text and unstructured data. Instead *of retrieving information* as we would from a structured dataset or a known set of search terms, text and data mining are concerned with *extracting information*, which is fundamentally different. Each of us is bombarded with information every day, much of it in numeric form, but the majority in text form. Mining the "gold" from data and text has become a very important task and a very big business. We can see an evolution of information from *data* to *information* to *knowledge* to *intelligence*. Organizations need ways to find and leverage the valuable intelligence in vast and unorganized collections of data and text, and that is what we will talk about in this chapter.

To solve this complex and challenging problem, we need a capable and scalable solution. R offers that and more. Before we get into specifics, let's examine exactly what data mining is (and is not). We might define data mining as the extraction of *predictive* (read valuable) information from (relatively) large datasets. To the extent possible, the extraction should be *automatic*, and the information should be predictive of something *valuable* to us. Implicit in this definition is that there is some *statistical methodology* that allows us to extract the information. There are many different data mining algorithms and tools, and it is somewhat of an idiosyncrasy to become enamored of a specific algorithm at the expense of understanding the results and their implications. Data mining tools typically include decision trees, various approaches to classification, the use of neural networks and machine learning, inducing rules or associations, reducing the dimensionality of data, and various approaches to clustering. Although there may be some overlap in purpose, data mining is not typically associated with data visualization, various queries of structured data, or data warehousing.

If you use statistics on a regular basis, you are probably already (at least) a fledgling data miner, perhaps without realizing it. For example, my doctoral dissertation involved the creation of subgroups from a sample of several thousand insurance agents on the basis of their responses to an autobiographical questionnaire with several hundred questions. I used principal components analysis to reduce the dimensionality of the biodata questionnaire, and then used hierarchical clustering of individual profiles to identify subgroups of agents who had various prior experiences in common. In support of the psychological adage that the best predictor of future behavior is past behavior, I found indeed that certain groups of agents were more likely than others to have both interest in and potential for becoming agency managers. I also found that other subgroups of agents might be very successful as agents, but were not interested in, nor likely to be good at, a management position with the company. At the time (back in the late 1970s), I had never heard of data mining, but the same techniques I used then are used in data mining today.

I did the data analysis for my dissertation using SAS running in batch mode on a mainframe IBM computer, performing a principal components analysis with varimax rotation that took from midnight to 6 a.m. As in many other areas, such as modern robust statistics, the techniques for data mining have often been conceptualized and developed long before the computing power and speed were available to make them feasible for use with today's incredibly large and loosely structured datasets.

Because data mining is a diverse field, there is not a single R package capable of all the data mining techniques that I mentioned. We will therefore have to pick and choose some illustrative applications. Progress, however, is being made toward the goal of a general-purpose R-based data mining package, and currently, the rattle package written by Graham Williams is the clear front-runner. The rattle package provides a graphical user interface for R and gives the user a point-and-click approach to data mining that works similarly to the way John Fox's R Commander package is used for standard statistical analyses. The rattle package must be installed as any other R package is, but when launched, provides its own Gnome-based interface (see Figure 16-1).

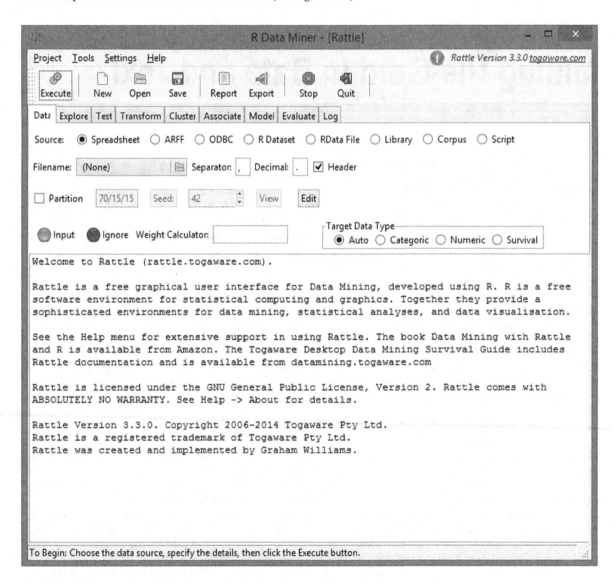

Figure 16-1. *The Rattle package running under R 3.1.2*

Beyond requiring the user to know how to install and launch an R package, Rattle does not require the user to be expert in the use of R, per se, but does assume familiarity with various approaches to data mining. The interface for rattle is laid out in such a way that it shows the progression of a typical data mining project, from loading data to various explorations, tests, transformations, and other manipulations of data to examining clusters or associations, choosing models, and then evaluating the models. We will return to rattle later in this chapter, comparing some of its output to that of R functions for the same analyses.

For now, we will use some of the built-in R functions for various data mining applications. Again, we will not delve into the depths of data mining or cover the entire set of procedures, but you will get a brief exposure to some of the most commonly used tools.

Recipe 16-1. Reducing the Dimensionality of Data
Problem

Data mining, as the name implies, seeks to discover information from unstructured data, when such information is not obvious. If the information were obvious, it would exist in a structured format. Though there are still challenges retrieving and analyzing data when the volume of structured data becomes very large, the challenge of *extracting* information from unstructured data is much greater. With large datasets consisting of multiple variables, it is often difficult to detect connections and patterns. It is also difficult to determine whether the interrelationships among the variables are such that we can reduce the dimensionality of the data to something more manageable and reasonable. In Recipe 16-1, you will learn how to use R for a principal components analysis (PCA). PCA and the closely related method of factor analysis have been used for many years, and have quickly demonstrated themselves to be useful in data mining applications.

Solution

For those interested in the history of statistics, factor analysis was developed in the early 1900s by British psychologist Charles Spearman, though most people are more likely to associate Spearman with the rank correlation coefficient. Spearman used factor analysis to support the theory that there is a general intelligence factor, g, which explains the relationships among different cognitive tests.

PCA is related to factor analysis, but PCA is more useful as a data *reduction* and descriptive statistical technique, whereas factor analysis is used to identify (exploratory) or verify (confirmatory) the presence of underlying unobserved (latent) variables called *factors* (in a completely different sense from a factor in an R data frame or table). In PCA, the primary goal is to find a relatively small number of "components" that can be used to summarize a set of variables without the loss of too much information. In PCA, the correlation matrix used to determine the number of components has 1's on the diagonal. This means that in PCA, we are seeking to explain (or account for) 100% of the variance (including the variance unique to each variable, the variance that is common among variables, and the error variance). In factor analysis, on the other hand, rather than unities (1's), the diagonal of the correlation matrix has "communalities" (that is, the variances shared in common with other variables), excluding the variance unique to each variable and error. For our current purposes, PCA will work fine.

Here are the 11 statements on a survey concerning the use of laptop computers in class (see Table 16-1). Statements marked with a superscripted *a* are negatively worded and reverse scored. My student Neal Herring and I developed the survey as part of a class project. The university where the survey was developed and administered has a requirement that students must have a laptop computer available for every class (ubiquitous computing requirement). We were interested in knowing whether students were in favor of this or not, and used Likert scaling to develop the survey from a much larger list of candidate items. The survey was administered to 65 students, with 61 complete cases (no missing data). Respondents rated their agreement or disagreement with each statement on a five-point response scale with 1 = *Strongly disagree*, 2 = *Disagree*, 3 = *Undecided*, 4 = *Agree*, and 5 = *Strongly agree*.

Table 16-1. *11-Item Scale of Students' Attitudes Toward Laptop Computers in Class*

Item	Statement
1	I pay better attention in class when I use my laptop for note-taking.
2[a]	Using a laptop in class does not increase my learning.
3[a]	It is a hassle to bring my laptop to class.
4	Using a laptop in the classroom makes me more efficient by reducing my study time outside of class.
5[a]	The weight of my laptop makes it hard to carry to class.
6	Using a laptop in class will help prepare me for a job or for advanced studies in my field.
7[a]	I find it distracting when others use their laptops in class.
8	Using a laptop in the classroom will make my grades higher.
9	I am more likely to attend classes where laptops are required or permitted.
10	Using a laptop in class makes class more interesting.
11	I think students should be permitted to use their laptops in all their classes.

Here are the first few rows of the dataset. See that in addition to item responses, we collected data on the respondents' sex, GPA, and class standing (1 = *freshman* to 4 = *senior*).

```
> head(laptops,3)
  id Item1 Item2 Item3 Item4 Item5 Item6 Item7 Item8 Item9 Item10 Item11 Sex
1  1     4     3     3     4     2     2     5     2     4      4      4   1
2  2     2     5     2     4     1     5     5     5     3      5      5   1
3  3     3     2     1     3     2     2     2     2     3      4      3   0
  Status  GPA
1      4 2.60
2      4 3.70
3      3 3.64
```

We will use the R package psych to find the principal components from the correlation matrix, as discussed earlier. We will also rotate the factor solution to a varimax (orthogonal rotation). The princomp() function in base R produces an unrotated solution, while the psych package allows rotation to several different solutions. As mentioned, the correlation matrix with 1's on the diagonal is used in PCA. We find that the 11 items are represented well by three principal components. The number of factors to extract is a bit of a controversial issue. A good rule of thumb is that any factor (or component) with an eigenvalue greater than 1 is a candidate for retention, but ultimately, the choice of the number of factors extracted often involves some subjective judgment based on the researcher's interpretation of the factor loadings. PCA uses an eigendecompostion of the square correlation or covariance matrix to produce uncorrelated (orthogonal) factors (represented by eigenvectors). The vector of eigenvalues represents the proportion of the total variance explained by each component, and of necessity, must sum to the number of original items as there are 1's on the diagonal of the correlation matrix. Let us convert our laptop survey item responses to a matrix and use the eigen() function to study the eigenvalues and eigenvectors. We simply subset the laptop data by selecting the 11 columns of item responses, and then use the cor function to produce our correlation matrix:

```
Items <- as.matrix(laptops[2:12])

> corMat <- cor(Items)
> corMat
            Item1     Item2     Item3     Item4      Item5     Item6     Item7
Item1  1.0000000 0.4231445 0.0511384 0.4632248 0.16643381 0.2183705 0.2547533
Item2  0.4231445 1.0000000 0.3432559 0.5255746 0.29104875 0.5608472 0.3984790
Item3  0.0511384 0.3432559 1.0000000 0.3186916 0.73461102 0.3065492 0.4264412
Item4  0.4632248 0.5255746 0.3186916 1.0000000 0.27198596 0.4911948 0.2528662
Item5  0.1664338 0.2910488 0.7346110 0.2719860 1.00000000 0.2568671 0.3633775
Item6  0.2183705 0.5608472 0.3065492 0.4911948 0.25686714 1.0000000 0.3198655
Item7  0.2547533 0.3984790 0.4264412 0.2528662 0.36337750 0.3198655 1.0000000
Item8  0.5146774 0.6398297 0.3175904 0.5751981 0.28702317 0.6802546 0.4252158
Item9  0.2245341 0.3846825 0.2548962 0.4542136 0.31866815 0.3373920 0.2738553
Item10 0.1668403 0.3887571 0.1902351 0.3770664 0.11772733 0.4267274 0.2666067
Item11 0.3247979 0.4227840 0.1262275 0.3682178 0.04660024 0.3849067 0.2605502
            Item8     Item9    Item10     Item11
Item1  0.5146774 0.2245341 0.1668403 0.32479793
Item2  0.6398297 0.3846825 0.3887571 0.42278399
Item3  0.3175904 0.2548962 0.1902351 0.12622747
Item4  0.5751981 0.4542136 0.3770664 0.36821778
Item5  0.2870232 0.3186681 0.1177273 0.04660024
Item6  0.6802546 0.3373920 0.4267274 0.38490670
Item7  0.4252158 0.2738553 0.2666067 0.26055018
Item8  1.0000000 0.3550858 0.4997211 0.45178437
Item9  0.3550858 1.0000000 0.5822105 0.30319201
Item10 0.4997211 0.5822105 1.0000000 0.44632809
Item11 0.4517844 0.3031920 0.4463281 1.00000000
```

Here are the eigenvalues and eigenvectors. We see that three components have eigenvalues greater than 1, as discussed earlier:

```
> eigen(corMat)
$values
 [1] 4.6681074 1.5408803 1.0270909 0.7942875 0.7335682 0.5896213 0.4394996
 [8] 0.4240592 0.3663119 0.2118143 0.2047596

$vectors
              [,1]        [,2]        [,3]         [,4]        [,5]        [,6]
 [1,] -0.2462244  0.22107473  0.58114752  0.447515243 -0.25136101 -0.02506555
 [2,] -0.3595942  0.06915554  0.16927109 -0.127637015  0.11454822 -0.07828474
 [3,] -0.2506909 -0.58829325 -0.04217347 -0.058873014  0.06611702  0.28137063
 [4,] -0.3390437  0.10370137  0.12366386  0.346201739  0.31534535  0.06486155
 [5,] -0.2334164 -0.60038263  0.03159179  0.203807163  0.04681617  0.19067543
 [6,] -0.3365345  0.08027461 -0.03125315 -0.419702484  0.47974475 -0.15637109
 [7,] -0.2681116 -0.23657580  0.10141271 -0.331553795 -0.65831805 -0.41855783
 [8,] -0.3854329  0.13565320  0.18437972 -0.175408233  0.14521421 -0.16420035
 [9,] -0.2898889  0.02672144 -0.50324733  0.497063117 -0.11046030 -0.19075495
[10,] -0.2947311  0.21566774 -0.55816283  0.008082903 -0.11482431 -0.09832977
[11,] -0.2705290  0.31780613 -0.07405850 -0.233659252 -0.32814554  0.77514458
```

```
              [,7]          [,8]          [,9]         [,10]          [,11]
 [1,]  -0.348837008 -0.006217720 -0.06016296 -0.32343802 -0.242340463
 [2,]   0.288701437 -0.788357274  0.28749655 -0.06086925  0.102473468
 [3,]   0.005269449  0.076511614  0.30203619 -0.02857311 -0.635889572
 [4,]   0.552818181  0.484383586  0.21064937 -0.03194788  0.223311571
 [5,]  -0.294511458 -0.091790248 -0.21930085 -0.03416408  0.599625476
 [6,]  -0.087058677  0.109537971 -0.49618779 -0.40890910 -0.124819321
 [7,]   0.281105085  0.218739682 -0.06533240 -0.05651648  0.095040577
 [8,]  -0.362175484  0.133992368  0.08834002  0.75050977 -0.005736071
 [9,]   0.159102079 -0.200051674 -0.42913546  0.22679185 -0.255745814
[10,]  -0.389241570  0.108029551  0.49528135 -0.31485066  0.156660772
[11,]   0.089435341  0.007522468 -0.20566674  0.06668854  0.049009955
```

Now, let us use the `principal` function in the `psych` package, extract three components, and rotate them to a varimax solution. The varimax solution rotates the axes of the factors while retaining their orthogonality. Rotations usually assist in the interpretation of factor or component loadings.

```
> pcaSolution <- principal(corMat, nfactors = 3, rotate = "varimax")
> pcaSolution$loadings
```

```
Loadings:
        RC1    RC2    RC3
Item1   0.835
Item2   0.675  0.284  0.323
Item3          0.897  0.133
Item4   0.631  0.225  0.347
Item5          0.894
Item6   0.519  0.246  0.458
Item7   0.356  0.529  0.161
Item8   0.753  0.237  0.366
Item9   0.126  0.252  0.758
Item10  0.185         0.872
Item11  0.502         0.495

                  RC1   RC2   RC3
SS loadings     2.830 2.205 2.201
Proportion Var  0.257 0.200 0.200
Cumulative Var  0.257 0.458 0.658
> pcaSolution$values
 [1] 4.6681074 1.5408803 1.0270909 0.7942875 0.7335682 0.5896213 0.4394996
 [8] 0.4240592 0.3663119 0.2118143 0.2047596
```

Although we are warned by some authorities that we should not attempt to interpret the factor loadings or the underlying variables in principal components, this is still routinely done by many researchers, including myself. The factor loadings are the correlations of the individual items with the rotated factors. The examination of the highest loadings shows that the first factor RC1 is related to learning and performance, the second factor RC2 is related to convenience and concentration, and the third factor RC3 is related to motivation and interest. Figure 16-2 is a structural diagram produced by the `fa.diagram` function in the `psych` package, which shows only the highest-loading items for each component. This diagram shows the latent structure and the loadings of the items on the factors.

```
> fa.diagram(pcaSolution)
```

Factor Analysis

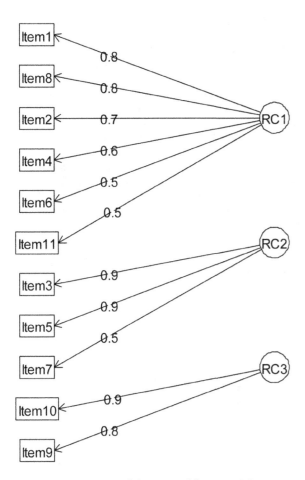

Figure 16-2. *Structural diagram of the rotated three-component solution*

Recipe 16-2. Finding Clusters of Individuals or Objects
Problem

We often have data that lend themselves to dimension-reduction techniques, as you saw in Recipe 16-1. The question then becomes, "So what? Now that I have a three-component solution explaining about 2/3 of the variance in students' attitudes toward the use of laptop computers in class, how can that help me?" Often the answer is that the components can be used to create groupings of the individuals or objects we are measuring. This is just as true of people as it is of animals or manufactured goods. These groupings may be useful for both empirical and theoretical purposes.

Solution

There are many ways to cluster objects or individuals. We will continue with our example and develop subgroups of the individuals who completed the survey described in Recipe 16-1. As with factor analysis, deciding on the appropriate number of subgroups or clusters is part art and part science. To use clustering, we need some metric. One of the most commonly used is the distance (difference) between objects or individuals. For my dissertation, I clustered insurance agents on the basis of the distance or similarity in their profiles based on factor scores. The distance matrix was used to develop a hierarchical clustering solution. Such solutions are commonly shown in a *dendrogram* (tree diagram). The algorithm starts with each object or individual as a separate entity, and then combines the objects or individuals into clusters by combining the two most similar clusters with each successive iteration until there is one cluster that contains all the objects or individuals. At some point, the researcher decides that there is a balance between the number of clusters and the specificity with which the differences between the clusters are interpretable. This is a classic problem in statistics, that of classifying objects or individuals.

We will continue with our laptop survey data, and extract the scores for each of the 61 students for whom we have complete data. Each person will have a score for each component. The principal function uses the traditional regression method by default to produce factor scores, and these are saved as an object in the PCA. We will then compute a 61 × 61 distance matrix, which will give us the distances between each pair of individuals. Next, we will use hierarchical clustering to group the individuals and plot the dendrogram to see what a good solution might be regarding the number of clusters to retain. Finally, on the basis of some demographic data and GPA, which were collected at the time the survey was administered, we will determine if the clusters of individuals are different from each other in any significant and meaningful ways. We will have to use the raw data rather than a correlation matrix for the purpose of producing the component scores, so we must run the PCA again using raw scores:

```
> Items <- as.data.frame(Items)
> head(Items)
  Item1 Item2 Item3 Item4 Item5 Item6 Item7 Item8 Item9 Item10 Item11
1     4     3     3     4     2     2     5     2     4      4      4
2     2     5     2     4     1     5     5     5     3      5      5
3     3     2     1     3     2     2     2     2     3      4      3
4     5     5     4     4     4     5     3     5     5      5      5
6     2     2     2     2     3     3     2     2     2      3      2
7     2     2     4     2     4     2     4     2     3      4      3
> pcaSolution <- principal(Items, nfactors = 3, rotate = "varimax")
> pcaScores <- pcaSolution$scores
> head(pcaScores)
          RC1         RC2        RC3
1   0.6070331 -0.1135057  0.2163606
2   1.5849366 -0.8321524  1.4991802
3  -0.2064288 -1.1233926  0.1235260
4   2.1816945  0.3399202  1.2432899
6  -0.6549734 -0.1209022 -0.6975968
7  -1.1428872  1.0286981  0.1273299
> distMat <- dist(as.matrix(pcaScores))
> hc <- hclust(distMat)
> plot(hc)
```

The dendrogram is shown in Figure 16-3. The numbers shown are the id numbers of the subjects, so that we can identify the cluster membership.

Cluster Dendrogram

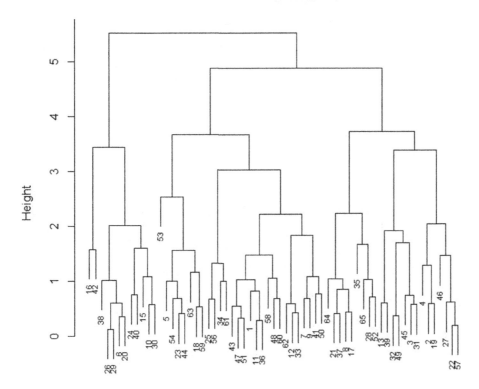

Figure 16-3. *Hierarchical clustering dendrogram for the laptop data*

The examination of the clustering reveals that three clusters may be a good starting point. To add rectangles surrounding the proposed clusters, use the rect.hclust function and specify the model name and the number of clusters as follows. After playing around with various solutions, I stuck with three clusters. The updated dendrogram appears in Figure 16-4.

```
> rect.hclust(hc, 3)
```

Cluster Dendrogram

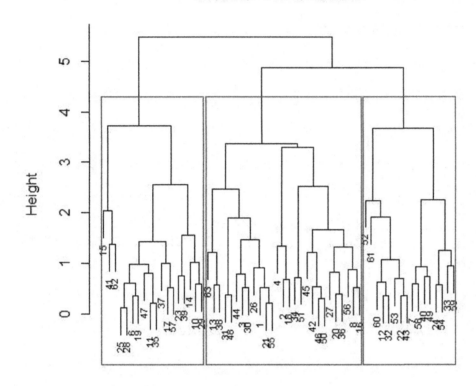

distMat
hclust (*, "complete")

Figure 16-4. *Updated dendrogram with clusters identified*

We can then save the cluster memberships using the cutree function, and append the cluster numbers to the laptop dataset as follows:

```
> cluster <- cutree(hc, 3)
> laptops <- cbind(laptops, cluster)
> head(laptops)
  id Item1 Item2 Item3 Item4 Item5 Item6 Item7 Item8 Item9 Item10 Item11 Sex
1  1     4     3     3     4     2     2     5     2     4      4      4   1
2  2     2     5     2     4     1     5     5     5     3      5      5   1
3  3     3     2     1     3     2     2     2     2     3      4      3   0
4  4     5     5     4     4     4     5     3     5     5      5      5   1
6  6     2     2     2     2     3     3     2     2     2      3      2   1
7  7     2     2     4     2     4     2     4     2     3      4      3   1
```

```
   Status  GPA cluster
1       4 2.60       1
2       4 3.70       1
3       3 3.64       1
4       3 3.00       1
6       3 2.80       2
7       3 3.60       3
```

Let's see if the clusters are useful for determining differences in GPA, sex, or status. We will use chi-square analyses for sex and status, and an analysis of variance for GPA. To make our analyses easier, we can attach the data frame so that we can call the variables directly. We determine from our chi-square tests that there are no significant associations between cluster membership and sex or status.

```
> attach(laptops)
> chisq.test(cluster, status)

        Pearson's Chi-squared test

data: cluster and status
X-squared = 8.9984, df = 6, p-value = 0.1737

Warning message:
In chisq.test(cluster, status) : Chi-squared approximation may be incorrect
> chisq.test(cluster, sex)

        Pearson's Chi-squared test

data: cluster and sex
X-squared = 3.0361, df = 2, p-value = 0.2191

Warning message:
In chisq.test(cluster, sex) : Chi-squared approximation may be incorrect

> oneway.test(GPA ~ cluster)
```

However, the three clusters of students have significantly different GPAs, and this would be worth examination to determine what other characteristics besides their attitudes toward laptops might be different for these groups of students. For those interested in knowing, my dissertation was based on a large-scale project for which I was a graduate research assistant at the University of Georgia. The principal investigator, the late Dr. William A. Owens, who was also my dissertation chair, was fond of saying, "You never really understand factor analysis until you do one by hand." I am sure that is true, but by the time I got around to learning it, there were computer programs available, so I never personally did a factor analysis by hand, even though I am quite positive Doc Owens was sure that it would have benefitted me greatly. Here are the results of the one-way test of cluster differences in GPA:

```
        One-way analysis of means (not assuming equal variances)

data:  GPA and cluster
F = 7.2605, num df = 2.000, denom df = 34.025, p-value = 0.002366
```

```
> aggregate(GPA ~ cluster, data = laptops, mean)
  cluster      GPA
1       1 2.992593
2       2 3.466111
3       3 3.193125
```

Recipe 16-3. Looking for Associations

Problem

Often, we use data mining to find previously undiscovered associations. In one common but most likely anecdotal example, a large retailer found that beer and diaper sales both increased on Friday. The story is told about other stores and other days, but the gist is always the same. By discovering a hitherto unknown association through data mining, the retailer was able to take advantage of the information and place displays of beer close to the disposable diapers. Presumably, because diapers come in large bundles, women asked their husbands to buy diapers, and with the weekend imminent, the man also decided to stock up on beer for the weekend. The location of beer and diapers in close proximity caused beer sales to increase. Moreover, the retailer could also charge full price for the beer because of the higher demand. This often-repeated tale is one of those that, as my former college provost said, "Should be true if it isn't."

Solution

In Recipe 16-3, we will examine association rule learning. In contrast to sequence mining, association rule mining is typically not concerned with the order of items within transactions or across transactions. One major consideration is that with large datasets, we often see associations that appear to be important, but are spurious and due to chance only. With large numbers of items, we must control the risk of interpreting such associations as meaningful or useful by establishing a higher standard for statistical soundness than we might employ in a controlled experiment.

Association analysis is especially useful for finding patterns and relationships to assist retailers, whether online or in stores. At a simplified level, imagine a shopping basket (or online shopping cart) that could have any of a collection of items in it. The aim of association rule analysis is to identify collections of items that appear together in multiple baskets. We can model a general association rule as

$$LHS \rightarrow RHS$$

where *LHS* is the left-hand side and *RHS* is the right-hand side of a grouping of items. Often we think of *LHS* as an *antecedent* and *RHS* as a *consequent*, but we are not strict on that because we are looking at correlations or associations rather than causation. Either the *LHS* or the *RHS* could consist of multiple items. For a grocery, a common pattern might be as follows:

$$cheese\ \&\ taco\ shells \rightarrow salsa$$

We will explore association rule analysis using the `arules` package. This package uses the apriori algorithm that is now part of IBM's proprietary SPSS Model Builder (formerly Clementine). Our example will be what is known as *basket analysis*. Think of a basket as representing a collection of items, such as a basket of shopping items, a panel of blood test results, a portfolio of stocks, or an assortment of medications prescribed to a patient. Each "basket" has a unique identifier, and the contents of the basket are the items contained in the basket listed in a column next to the identifier, which could literally be a shopping basket, but could also be a particular customer, a patient, an online shopping cart, or some other unit of analysis. The contents of the baskets are the target variable in the association rule analysis.

In association rule analysis, the search heuristic looks for patterns of repeated items, such as the examples given earlier. We want to find combinations of items that are "frequent enough" and "interesting enough" to control the number of association rules to something manageable. The two primary measures used in association analysis are *support* and *confidence*. Support is expressed as some minimum percentage of the time two or more items appear together. It must typically be set fairly low, because obvious transactions that appear together frequently (say *chips* and *dip*) are of less interest than unusual ones (such as *beer* and *diapers*). Confidence is also expressed as a proportion, but more formally, it is a conditional probability; that is, the probability of *RHS* given *LHS*, which equals

$$\text{Confidence}(LHS \rightarrow RHS) = P(RHS \mid LHS) = \frac{P(RHS \cap LHS)}{P(LHS)}$$

Another measure of interest is *lift*, which we can define as the improvement in the occurrence of *RHS* given the *LHS*. Lift is the ratio of the conditional probability of *RHS* given *LHS* and the unconditional probability of *RHS*:

$$\text{Lift}(LHS \rightarrow RHS) = \frac{P(RHS \mid LHS)}{P(RHS)} = \frac{\text{Confidence}(LHS \rightarrow RHS)}{\text{Support}(RHS)}$$

Let us use the arules package and perform an apriori rules analysis of the dataset called "marketBasket.csv." The data consist of 16 different items that might be bought at a grocery store, and 8 separate transactions (baskets). The data are as follows:

```
> basket <- read.csv("marketBasket.csv")
> basket
   basket     item
1       1    bread
2       1   butter
3       1     eggs
4       1     milk
5       2  bologna
6       2    bread
7       2   cheese
8       2    chips
9       2     mayo
10      2     soda
11      3  bananas
12      3    bread
13      3   butter
14      3   cheese
15      3  oranges
16      4     buns
17      4    chips
18      4  hotdogs
19      4  mustard
20      4     soda
21      5     buns
22      5    chips
23      5  hotdogs
24      5  mustard
25      5  pickles
26      5     soda
27      6    bread
28      6   butter
29      6   cereal
30      6     eggs
```

```
31      6     milk
32      7 bananas
33      7  cereal
34      7    eggs
35      7    milk
36      7 oranges
37      8 bologna
38      8   bread
39      8    buns
40      8  cheese
41      8   chips
42      8 hotdogs
43      8    mayo
44      8 mustard
45      8    soda
```

To use the apriori function in the arules package, we must first convert the dataset into a transaction data structure:

```
> library(arules)
> newBasket <- new.env()
> newBasket <- as(split(basket$item, basket$basket), "transactions")
> newBasket
transactions in sparse format with
 8 transactions (rows) and
 16 items (columns)
```

Now, we can use the apriori function to find the associations among the items in the eight shopping baskets. We will save the rules as an object we can query for additional information. We specify the parameters for the association rule mining as a list. We will look for rules that have support of at least .20, confidence of .80, and minimum length of 2, which means that there must be both a *LHS* and a *RHS*. The default is a length of 1, and because *RHS* must always have 1 (and only 1) item, this would result in a rule in which the *LHS* is empty, as in {empty} → {*beer*}.

```
> rules <- apriori(newBasket, parameter = list(supp = .2, conf = .8, minlen = 2, target = "rules"))

parameter specification:
 confidence minval smax arem  aval originalSupport support minlen maxlen target
        0.8    0.1    1 none FALSE            TRUE     0.2      1     10  rules
   ext
 FALSE

algorithmic control:
 filter tree heap memopt load sort verbose
    0.1 TRUE TRUE  FALSE TRUE    2    TRUE

Warning in apriori(newBasket, parameter = list(supp = 0.2, conf = 0.8, target = "rules")) :
  You chose a very low absolute support count of 1. You might run out of memory! Increase minimum
support.
```

```
apriori - find association rules with the apriori algorithm
version 4.21 (2004.05.09)         (c) 1996-2004    Christian Borgelt
set item appearances ...[0 item(s)] done [0.00s].
set transactions ...[16 item(s), 8 transaction(s)] done [0.00s].
sorting and recoding items ... [15 item(s)] done [0.00s].
creating transaction tree ... done [0.00s].
checking subsets of size 1 2 3 4 5 6 done [0.00s].
writing ... [246 rule(s)] done [0.00s].
creating S4 object ... done [0.00s].
```

Although we were warned about potential memory limitations, none materialized. The algorithm located 246 rules fitting our criteria. As a point of reference, if we had accepted the defaults, we would have had more than 2,000 rules to consider.

```
> rules
set of 246 rules
> summary(rules)
set of 246 rules

rule length distribution (lhs + rhs):sizes
 2  3  4  5  6
32 89 84 35  6

   Min. 1st Qu.  Median    Mean 3rd Qu.    Max.
  2.000   3.000   4.000   3.569   4.000   6.000

summary of quality measures:
     support        confidence       lift
 Min.   :0.2500  Min.   :1    Min.   :1.600
 1st Qu.:0.2500  1st Qu.:1    1st Qu.:2.000
 Median :0.2500  Median :1    Median :2.667
 Mean   :0.2866  Mean   :1    Mean   :2.582
 3rd Qu.:0.3750  3rd Qu.:1    3rd Qu.:2.667
 Max.   :0.5000  Max.   :1    Max.   :4.000

mining info:
      data ntransactions support confidence
 newBasket            8     0.2        0.8
```

We can use inspect() to find the rules with high confidence (or support or lift) as follows:

```
> inspect(head(sort(rules, by = "confidence")
+ )
+ )
  lhs           rhs         support confidence      lift
1 {cereal}  => {eggs}        0.250          1 2.666667
2 {cereal}  => {milk}        0.250          1 2.666667
3 {bananas} => {oranges}     0.250          1 4.000000
4 {oranges} => {bananas}     0.250          1 4.000000
5 {butter}  => {bread}       0.375          1 1.600000
6 {eggs}    => {milk}        0.375          1 2.666667
```

Let us use the `rattle` package for this same application to illustrate how to do the association rule analysis. Install the `rattle` package, and then load and launch it using the following commands:

```
> install.packages("rattle")
> library("rattle")
Rattle: A free graphical interface for data mining with R.
Version 3.3.0 Copyright (c) 2006-2014 Togaware Pty Ltd.
Type 'rattle()' to shake, rattle, and roll your data.
> rattle()
```

The Rattle GUI() opens in a new window. Every operation in Rattle requires that you execute it. For example, if you choose a dataset but do not click the Execute icon, the data will not be loaded into Rattle. We will load the `marketBasket` data and then use Rattle to "shake, rattle, and roll" the data, as the package advertises. Several example datasets come with Rattle, the most complex of which is the "weather" dataset. Rattle uses the `arules` package and the apriori function for the association rule analysis, just as we have done. Rattle will install any additional packages it requires on the fly, which is a nice feature, but which can be disconcerting when you first begin using Rattle.

We will load the CSV version of our basket dataset(), and compare the output of Rattle to that of the `arules` package. There are eight data types supported by Rattle (see Figure 16-1). We will use the "spreadsheet" format for our source, which is really CSV.

1. Click the little folder icon, navigate to the required directory, and then click the file name to open the data (see Figure 16-5).

2. Click Open and then click Execute in the Rattle icon bar (or use the shortcut F2). Failure to execute will mean the file will not be loaded.

3. In the Data tab, change the role of the basket variable to *Ident* and the role of the item variable to *Target.*

4. Rattle can be used for two kinds of association rule analysis. The simpler of these is market basket analysis, and it uses only two variables—an identifier and a list of items, as we have been doing. The second approach is to use all the input variables to find associations. To select basket analysis, check the box labeled Baskets on the Associate tab in Rattle. Remember that clicking Open will not actually load the data. You must also click Execute to bring the dataset into R.

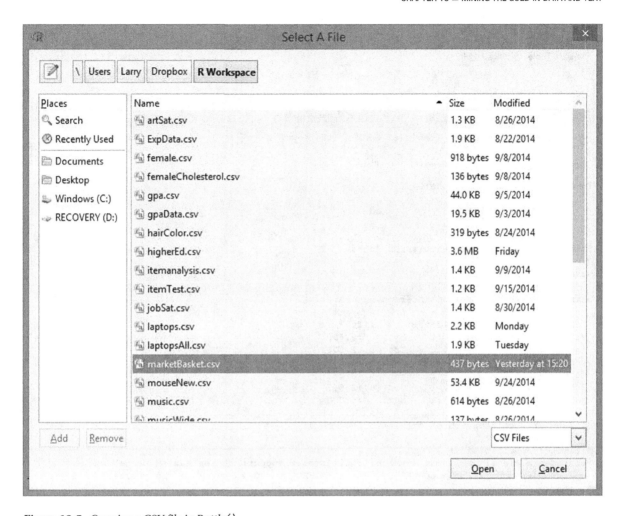

Figure 16-5. *Opening a CSV file in Rattle( )*

Let us use the same settings in Rattle that we did in `arules`, and compare the results. Because Rattle uses the same R packages and functions we used, our results should be the same. Remember to click Execute after specifying a basket analysis and changing the settings in Rattle to match the ones we used in `arules` (see Figure 16-6).

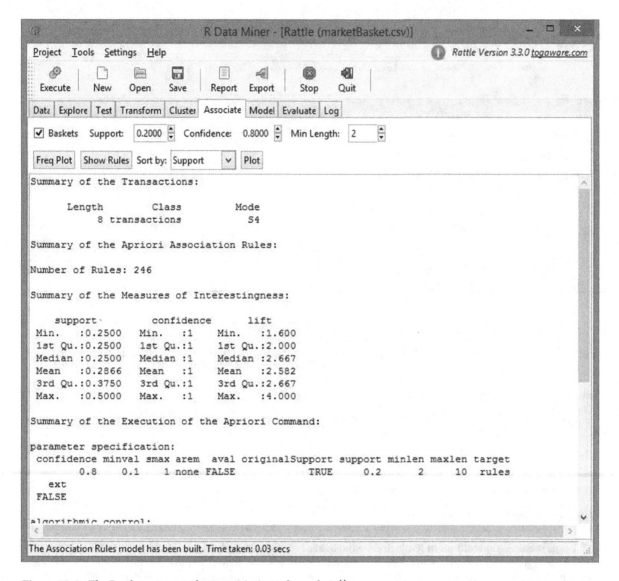

Figure 16-6. *The Rattle summary of our association rule analysis( )*

We did get the same results with Rattle that we did using the R functions, but Rattle packaged the results very nicely and added features to view plots and sort the rules, which we had to do manually in R.

Recipe 16-4. Mining Text: A Brief Introduction

You learned about working with strings earlier in this book in Recipes 2-3 and 5-2, and you were briefly introduced to the concept of text mining. In Recipe 16-4, we will cover text mining in enough depth to get you started mining text on your own. One of the "coolest" things you can do with text mining, at least in my opinion, is to produce "word clouds," so we will do that as well.

Problem

Written and spoken words are everywhere, and are being recorded, stored, and mined with greater frequency than ever before. Although the focus of data mining is on finding patterns, associations, trends, and differences between and among data values that are primarily numeric, the greater majority of the information in the world is text rather than numbers, as we discussed previously.

Text mining is used in a number of different fields, just as data mining is, and there are many ways to go about mining text. Finding patterns and associations among words becomes very important when such patterns and associations help solve practical problems. I am a *quantitative* researcher by training and preference, but *qualitative* research, which uses narratives the way quantitative researchers use numbers, is growing both in importance and popularity. Interpreting the results of interviews, written narratives, and other text-based data is becoming pervasive, and qualitative research is seen by many, if not most, as being on a par with quantitative research. In fact, many research projects now combine both quantitative and qualitative methods in what is known as *mixed-methods research*.

As with all of the techniques discussed in this chapter, there are elements of judgment and subjectivity in text mining, just as in data mining. Those analysts with keener (and often more intuitive) judgment are obviously more effective than those who simply follow heuristic rules dogmatically. At present, computers are still relatively "dumb" when it comes to understanding what words spoken or written by humans really mean. For example, what would a computer program "think" of the sentence, "Time flies like an arrow."? Most humans know intuitively that this is a simile, and that we are saying that time passes swiftly, just as an arrow flies (relatively) swiftly. But a computer could just as easily take this as an instruction to time the flight speed of houseflies in the same way that it would time the speed of the flight of an arrow.

As of today, computers don't really have much of a sense of humor, or much intuition, though advances are being made every day, and the day may soon come when a computer program can pass the Turing test, in which a human judge cannot determine reliably, whether he or she is talking to a computer or to another human. Alan Turing posed the question in 1950 as to whether machines can "think," though his question was presaged by Descartes in his musings about automata centuries earlier.

The year 2012 marked the 100th anniversary of Turing's birth, and as of today, the Turing test has not been passed with anything better than a 33% rate of a computer program convincing a human that it is another human in conversations facilitated by keyboards and computer monitors rather than face-to-face communication. Text mining combines analytics with natural language processing (NLP). Until the 1980s, NLP systems were based mostly on human-supplied rules, but the introduction of machine learning algorithms revolutionized the field of NLP. A discussion of NLP and statistical machine learning to the contrary notwithstanding, text mining is still a viable and valuable endeavor, assisted by computer programs that allow us to mine the depths of text in ways we only imagined previously.

Solution

The tm package in R is very widely used for text mining. There are many other tools used by text miners as well. We will experiment with several of these. In text mining, we are interested in mining one or more collections of words, or *corpora*. Inspection of Rattle's interface (see Figure 16-1) reveals that two of the input types supported are corpus and script. A *corpus* is a collection of written works. In text mining, some of the more mundane tasks are removing "stop words" and stripping suffixes from word stems so that we can mine the corpus more efficiently.

Text mining is also known as *text data mining* or *text analytics*. Just as in data mining in general, we are looking for information that is *novel*, *interesting*, or *relevant*. We might categorize or cluster text, analyze sentiment, summarize documents, or examine patterns or relationships between entities.

We will use the tm package to mine a web post about Clemson University and CUDA (Compute Unified Device Architecture), which is a parallel computing platform developed by NVIDIA and implemented by using the graphics processing units (GPUs produced by NVIDIA). Clemson (where I teach part-time) was recently named a CUDA teaching institution and a CUDA research center. The announcement was published online, and I simply copied the text from the web page into a text editor and saved it. The article is used with the permission of its author, Paul Alongi.

To get the text into shape for mining, we must read the text into R, convert it to a character vector, and then convert the vector to a corpus. Next, we need to perform a little "surgery" on the corpus, converting everything to lowercase, removing punctuation and numbers, and removing the common English "stop words" and extra spaces. We can use the tm_map function to create and revise the corpus. First, we need the tm package and the SnowballC package:

```
library(tm)
library(SnowballC)
text.corpus <- readLines("CLEMSON.txt")
text.corpus <- Corpus(VectorSource(text.corpus))
text.corpus <- tm_map(text.corpus, tolower)
text.corpus <- tm_map(text.corpus, removePunctuation)
text.corpus <- tm_map(text.corpus, removeNumbers)
text.corpus <- tm_map(text.corpus, removeWords, stopwords("english"))
text.corpus <- tm_map(text.corpus, stripWhitespace)
text.corpus <- tm_map(text.corpus, PlainTextDocument)
```

Now, we can "mine" our corpus, which is seen by tm as 19 separate documents (these are the paragraphs in the article), by finding frequent terms and associations in a fashion similar to what we did in data mining. To do this, we need to convert the corpus to a *term document matrix*, and then we can find the frequent terms and associations as follows. We set the correlation limit, which is a minimum value, to 0.5. We specify the search term and the text document matrix as well.

```
> tdm <- TermDocumentMatrix(text.corpus)
> findAssocs(x = tdm, term = "clemson", corlimit = 0.5)
              clemson
university        0.66
advancing         0.51
announced         0.51
commitment        0.51
david             0.51
distinguished     0.51
educate           0.51
education         0.51
forward           0.51
generations       0.51
inventor          0.51
leader            0.51
look              0.51
luebke            0.51
monday            0.51
senior            0.51
state             0.51
using             0.51
visual            0.51
working           0.51
world             0.51
> findFreqTerms(x = tdm, lowfreq = 10, highfreq = Inf)
[1] "clemson"   "computing" "cuda"      "nvidia"
```

If we wanted to continue mining, we might want to "stem" the corpus, which will truncate words to a common stem, so that "compute," "computer," and "computing" would all become "comput." This operation requires the SnowballC package we loaded earlier. We can then use the tm_map function in the tm package to find the specific stem elements and where they occur within the corpus. We examine only the first few of these terms, namely the first 20 rows and the first 5 columns (which correspond to the first 5 paragraphs of the article).

```
> myStems <- tm_map(text.corpus, stemDocument)
> myCorpus <- TermDocumentMatrix(myStems)
> inspect(myCorpus[1:20,1:5])
<<TermDocumentMatrix (terms: 20, documents: 5)>>
Non-/sparse entries: 3/97
Sparsity           : 97%
Maximal term length: 7
Weighting          : term frequency (tf)
```

Terms	Docs character(0)	character(0)	character(0)	character(0)	character(0)
acceler	0	0	0	1	0
access	0	0	0	0	0
accord	0	0	0	0	0
across	0	0	0	0	0
advanc	0	0	0	0	0
aim	0	0	0	0	0
air	0	0	0	0	0
alreadi	0	0	0	0	0
also	0	0	0	0	0
analyz	0	0	0	0	0
announc	1	0	0	0	0
applic	0	0	0	0	0
array	0	0	0	0	0
assist	0	0	0	0	0
automot	0	0	0	0	0
base	0	0	0	0	0
benefit	0	0	0	0	0
better	0	1	0	0	0
build	0	0	0	0	0
calcul	0	0	0	0	0

I mentioned word *clouds* earlier. The wordcloud package will create a nice-looking word cloud from a corpus or a text document. Figure 16-7 shows a word cloud (also known as a *tag cloud*) from the text we have been mining.

```
> library(wordcloud)
> wordcloud(text.corpus)
```

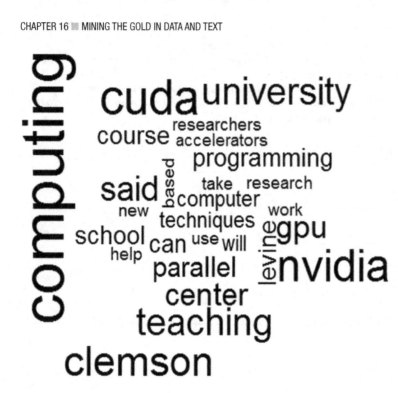

Figure 16-7. *A word cloud developed from our corpus*

Index

■ W, X

■ Y, Z

Get the eBook for only $10!

Now you can take the weightless companion with you anywhere, anytime. Your purchase of this book entitles you to 3 electronic versions for only $10.

This Apress title will prove so indispensible that you'll want to carry it with you everywhere, which is why we are offering the eBook in **3 formats** for only $10 if you have already purchased the print book.

Convenient and fully searchable, the PDF version enables you to easily find and copy code—or perform examples by quickly toggling between instructions and applications. The MOBI format is ideal for your Kindle, while the ePUB can be utilized on a variety of mobile devices.

Go to www.apress.com/promo/tendollars to purchase your companion eBook.